STEPS TOWARDS THE SUMMIT

The stages of human evolution.

A Study by Thomas Sigley

Steps towards the Summit is a revised version of *From Microbes to Man.* The latter is still available at Amazon, along with the digital (Kindle) edition.

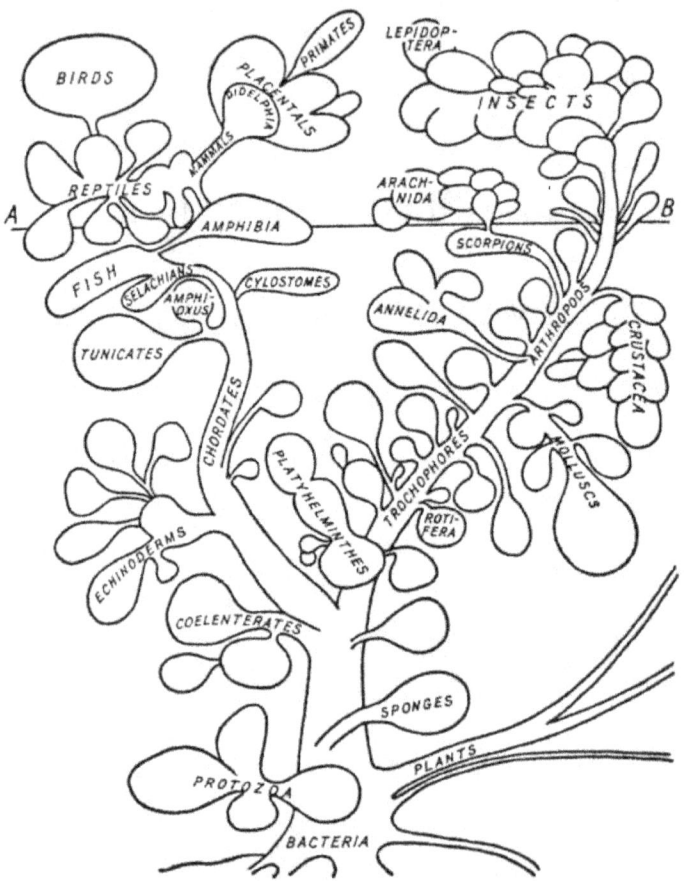

DIAGRAM 2. *The 'Tree of Life' after Cuénot. On this diagram each principal lobe (or bunch) represents a grade at least as important (morphologically and quantitatively) as that of the whole of the Mammalia taken together. Below the line AB, the forms are aquatic; above it they live on land.*

CONTENTS

Foreword

To discover the origin of life is a project that has engaged scientists, unsuccessfully, over the centuries. However, their studies on how lower forms of life have evolved into higher forms have been rewarded with some amazing results. The stages of evolution throughout history, leading to our own *Homo sapiens* species, presents itself to me as a most intriguing subject.

My interest was first aroused by the need to explain evolution to High School students, who found some explanations of it a stumbling block to their belief in God. This interest led me to read a lot of articles on the internet. This was followed by my reading three memorable books about evolution.

***Becoming Human* by** Ian Tattersall[1] was the first. He speaks of the "emergent quality" of Homo sapiens. On page 189 he explains "emergent" as follows: "The classic example of such a quality is water, whose remarkable characteristics, so essential to life, are entirely unpredicted by the hydrogen and oxygen atoms alone."

The second book I found inspiring was *The Signature in the Cell,*

---

[1] Curator of the American Museum of Natural History. Published in 1998.

by Stephen Meyer. It was "The Book of the Year" chosen, in 2009, by the Literary Supplement of Time magazine. Meyer maintains that science defines its object as concerned with matter and energy, but the discovery of the DNA genetic code makes it necessary to include "information" in its definition, a move that can hardly be refused, and which would open up the definition of science to the inclusion of a search for the explanation of the origin of the DNA code. He claims that only to have accepted the existence of the code without recognizing it as a product of intelligence is a fundamental mistake.

The next book that impressed me deeply is *The Language of God*, written by Francis S. Collins, the man in charge of the project to map the human genome. His position is what is called TE (Theistic Evolution). He maintains that this is "the dominant position of serious biologists who are also serious believers." Theistic evolution is the name given to the way of thinking that, based on faith or philosophical reasoning, regards the existence of God as indisputable. It looks to science to explore the natural phenomena in the universe and to unravel the riddles in biological history.[2]

This "TE" or Theistic Evolution is also the way of thinking closest to that of the Catholic Church. [3] It is less commonly called " Evolutionary Creationism". Fr. George Coyne, the Chief Vatican astronomer, stated forcefully, " Intelligent Design (ID) isn't science, even though it pretends to be." Chapter 13,p.103 of this book

---

[2] *The Language of God pp.199,200 &201*
[3] Wikipedia: Theistic Evolution and the Roman Catholic Church

explains why ID is unacceptable.

The more I read, the more conscious I became of how superficial my knowledge was, but Jacques Monod (the discoverer of Messenger RNA), in his [4] Herbert Spencer Lecture in 1996, said 'Another curious feature of the theory of evolution is that everybody thinks he understands it.'

Something I have noticed in my enquiries about evolution is the frequent appearance in newspapers and magazines of the subject and its related DNA. I think this shows that a lot of people besides myself are interested in the findings of modern science relating to evolution and heredity.

In the Queensland *Courier Mail* (2011) there was a statement "ID to be taught in Queensland schools under the National Curriculum" What interested me especially was the over 500 mostly negative letters that followed the announcement. I read most of them and noticed that many of the writers didn't have a good grasp of the problem, so it is my hope that this brief, overall picture of the evolution of modern man may shed some light on the subject.

This book will take us up to the point where Homo sapiens or we modern people, with intelligence characterizing us, appeared in Africa about 200,000 years ago.

_____

[4] Quoted on p.18, *The Selfish Gene,* by Richard Dawkins

The next step in evolution will show how intelligent man relates to God. It is treated in the sequel to this book. Its title is *The Best of Life*. Because of the nature of the problem presented, this book is predominately a religious book based on the Judeo/Christian Bible.

It shows that man is destined for a higher calling than the calling to increase and multiply and to subdue the earth. This will be the summit of evolution.

April, 2014

Chapter 1

The Big Bang

Richard Dawkins introduces his ideas about the origins of life with the words: " Complex atoms have been being formed in stars all over the universe, ever since soon after the 'big bang' which, according to the prevailing theory, initiated the universe. This is originally where the elements on our world came from." [5]

The 'big bang' is often regarded by ordinary people as a rather amusing theory, but it now seems to have been proved beyond doubt. The theory was rejected at first, but its supporters maintained that, if it had indeed occurred, the explosion would have left a cosmic microwave background (CMB) throughout the universe.[6] Two scientists Amo Penzias and Robert Wilson, of Bell Laboratories in New Jersey, won the Nobel Prize for their discovery of this radiation in 1965. They stumbled onto their conclusions when they were trying to create the world's most sensitive radio antenna. No matter where they pointed their antenna they encountered a strange static indicative of heat.   Their antenna was so sensitive that, at one stage they even theorized that it might be due to the heat of bird droppings in the sky! Gradually, they realized that they were hearing

---

[5] *The Selfish Gene,* Oxford University Press. 30th Edition, 2009. p.13

[6] Google: *Big Bang telescope probe catches first images of our universe.*

the microwave background left over from the birth of the universe.

Astronomers had accepted the theory from the start because they had found what they regarded as evidence that the universe was really expanding just as the sparks of a huge fireworks display radiate away from their source and seem to fade into the distance. It was the American-born astronomer Edwin Powell Hubble (1889-1953) who first discovered by telescopic observation that the Milky Way was not the whole universe, but there were other galaxies beyond it. He also discovered, by observing light-wave lengths, that the galaxies were moving away from one another at a rate constant to the distance between them. He formulated the Hubble Law in 1929, which was one of the ways scientists have used to determine the age of the universe. (estimated now as 14 billion years) He had been elected to the National Academy of Science in 1927, [7] the youngest person to have received this honor. His sudden death in 1953 is said to have prevented his receiving the Nobel Prize in 1954).

This theory appealed to one of the world's greatest scientists, Albert Einstein (1879 –1965). He had published his seminal papers proving the existence of the atom, the theory of relativity, and had described quantum mechanics by the time he was 26 years old. His theory of relativity left him dissatisfied because it was hard to reconcile it with the static and immobile universe that he accepted as a premise. When Hubble had proved his point, Einstein accepted it in 1931 with no reservations, saying, " This is the most

---

[7]   Cf. Google ,Edwin Powell Hubble, National Academy of Science.

beautiful and satisfactory explanations of creation that I have ever heard." He referred to his former acceptance of a static universe as " the greatest blunder of my scientific career." He had accepted it grudgingly because the universe his theory of relativity revealed seemed too far-fetched to be true.

The discoveries of scientists like Hubble, Penzias, Wilson and Albert Einstein show that the universe had a beginning. The silence and darkness of the void had been shattered by an explosion of unimaginable proportions[8]. The laws of nature came into being for curious intelligent people to discover, unravel and utilize. Time and space came into being on that fateful day and one of the effects of the 'big bang' was the formation of the stars and galaxies and, in due time. the formation of the earth.

Stephen Hawking, the English theoretical physicist and cosmologist, a scientist not noted for his religiosity, remarks in his *A Brief History of Time p.144* on the way the universe seems to have been tuned for biological life, " It would be difficult to explain why the universe should have begun in just this way, except as the act of a God who intended to create beings like us." [9]

---

[8] Japan Times 22/10/'10 : The most distant galaxy known in the universe is called UDFy-38135539. A light that appears to be receding from the observer, shifts more towards the red part of the optical spectrum. This galaxy's red-shift was 8.6 out of a possible 10. The light from this galaxy was emitted when the cosmos was only 600 million years old. It has taken 13.1 billion years to arrive, traveling at 300,000 km per second.

[9] In a more recent book, *The Grand Design,* remarking on the 1992 discovery of another planetary system similar to our own, Hawking states that the lucky combination of earth-sun distance and solar mass is less remarkable ,thus, less compelling as evidence that the earth was carefully designed just to please us.

Chapter 2

The Origin of Biological Species

Nobody was around to hear the 'big bang' or see the fireworks that accompanied the explosion,[10] but the photographs taken by the modern Hubble Telescope show pictures of cosmic explosions that occurred hundreds of light years ago, but which have just reached us now. Their explosive force of light and color give us hints of what the origin of the universe would have been like. Modern man, with his intelligence and curiosity, didn't appear on the scene until comparatively recently (200,000 years ago), but, once he got on his feet, it wasn't long before he began to make enquiries about how it all began.

The age of the earth was once a matter of controversy, but modern radiometric dating, recognized by the specialists, indicates that the earth began (coalesced from cosmic materials into a planet circling the sun) **4,750 million years ago**. Geologists had been researching the fossil record for about two centuries before the present day, and they had unearthed quite a lot of evidence allowing them to draw the conclusion that life forms on earth had been many and varied and had existed over vast tracts of time. However, the scholars were far from being able to answer the questions in their mind about the origins of man himself. Richard Dawkins quotes[11]

---

[10].Cf.Google. Photographs taken by the modern Hubble Telescope.

[11] Ch.1/p.1 *The Selfish Gene*

the zoologist G.G. Simpson: ' All attempts to answer that question before 1859 are worthless and we will be better off if we ignore them completely.'

1859 is the year Charles Darwin published his book about evolution " ***The Origin of the Species by Natural Selection.*** " This became one of the most influential books ever written. Scientists have based their research about the origin of man on the theory of evolution as expounded by Darwin.

Darwin was born in Shrewsbury, England, in 1809. He entered Cambridge University, intending to become a clergyman, but changed his mind and began to follow his natural scientific bent. As an avid naturalist, he came to believe that all life evolved from the first forms that made their appearance soon after the birth of the earth. He made the verification of this hypothesis the purpose of his life.

In 1831 he joined a scientific expedition that sailed on " The Beagle" to circle the world. On the way, they visited the isolated Galapagos Islands in the Pacific Ocean, where he found many species of animals that varied slightly from those of a similar lineage in South America. For instances, on the journey, he had collected samples of finches that differed from one another and from those of South America by the size and shapes of their beaks. These birds were all new species of the same genus and Darwin concluded that minute genetic changes, isolation on their respective island homes and the passing of time had brought their diversification about. He regarded this pattern as the origin of biological species.

The author of an article on Darwin pointed out, [12]'This explanation seemed more logical than assuming the bird types had each been created separately and placed in the various Galapagos Islands individually by the Creator of the universe."

The following is an outline of what we could call "Classical Darwinism".

What Darwin suggested was that a freak or **"random" genetic difference** could give an individual an advantage for better survival in a particular environment. This random genetic difference would be handed down hereditarily and gradually the better endowed individual's descendants would flourish while those without the advantage-giving characteristic would eventually die out.

The trouble with this new idea was that it ran contrary to the prevailing religious beliefs about God's creation of life forms. Darwin's own wife was a devout believer and he was a man who was careful with his words, not wanting to offend people. However, it seems that his convictions about his theory of evolution never wavered in spite of the fact that its seeming contradiction of his religious beliefs gave rise to much soul searching thought. By the time he got around to publishing his *Origin of the Species*, he had apparently found ways to reconcile his belief in God with his scientific theories.[13]

Those who believe in God may feel an aversion to the idea of evolution being dependent on <u>random</u> or chance genetic mutations

---

[12] Cf. *National Geographic* (Vol.206.No.5, November, 2004) "*Was Darwin Wrong*" by David Quammen.
[13] P.108 of this book.

in an organism. How could the Creator take such a chance as to found the evolution of life forms on random chance?

The question is unnecessary. The fact that the mutations are random doesn't mean that the whole edifice of evolution is based completely on chance. It could be expressed in another way: Nature has arranged that a continuous supply (About 60 per one generation) of genetic mutations be available, knowing that advantageous ones would make the recipient more fit to attain its potential to exist optimally in the environment in which it exists.

We could exemplify this by Darwin's finches in the Galapagos Islands. This natural process of selection is all that is required. There is no need for further supernatural intervention, just as there is no need for supernatural intervention to ensure that the laws of Newtonian gravity have their effect. (The hypothesis of Intelligent Design (ID) claims that supernatural intervention is necessary in the case of very complex organisms that couldn't have evolved piecemeal, but this so-called premise undermines the general law and is not accepted by the majority of scientists, even those who believe in the existence of God.)

However, what becomes evident is the fact that evolution, though inevitable, would have sometimes been delayed until a selectable mutation happened to come along. For this reason, it seems likely that other regulatory factors might have been involved in the evolutionary process in addition to the random mutations that Darwin tied his theory to.

When a random advantageous genetic modification is inherited in a species, those who have it share their living space, to a gradually receding extent, with those who haven't inherited the quality. In other words, the better endowed specimens gradually branch away from the parent stock, but still coexist with it. This is called "cladogenesis" (cladus = a branch or sprout). A more thorough degree of evolution is called "anagenesis" or "phyletic change." This means that, over time, a **whole population** becomes different to its ancestral population, and a new species name has to be assigned. This is called **speciation** and implies that the new group can't interbreed with the ancestral group. This is called **divergence.**

A few more notes on Darwinism could be added here. This is pointed out by David Quammen in the National Geographic Magazine cited previously. Darwin's work involved four separate branches of observation: *Biogeography; Paleontology; Embryology;* and *Morphology.*

Biogeography researches the reasons why similar species occur in different locations. It concludes that they have come from common ancestors at some time and place when they did share the same location.

Paleontology studies the vertical columns of geological strata to discover fossils that are a record of when certain life-forms first appeared, how they gradually changed and how and when they became extinct.

Embryology studies why, for instance, the embryo of mammals contain vestiges that resemble the embryo of reptiles. Darwin held that embryos can reveal the structures of their forebears and show the evolutionary history of any species.

Morphology studies the physical characteristics that a

species might share with another species e.g. fish, birds and animals all have vertebrae and are thus categorized as vertebrates, although they seem to differ radically The number of physical characteristics a species shares with another species indicates how recently those species have diverged from each other.

Present-day scientists have many advantages that Darwin didn't enjoy. Genetics has advanced amazingly since Darwin's day. Gregor Mendel, an Austrian monk, had discovered the fundamentals of genetics in Darwin's time, but his ideas had been ignored and were unknown to Darwin himself. By experiments, Mendel demonstrated that inheritance of units of information (He didn't know about genes, but his observations showed that something akin to genes must exist) must have been the driving force behind inherited traits. When scientists merged Darwinian evolutionary ideas with genomics, the histories of the various species and their mutual relationships were able to be understood with much more clarity and certainty.

In order to establish the fact that all life forms have evolved from the very first life forms to appear on the earth, Darwin realized that huge expanses of time would have been needed because evolution due to random genetic changes would be a very gradual process.

In the next chapter, we will glance at how much time has been available for the changes in flora and fauna that we find on earth.

Chapter 3

Counting the Days in Geology

The figures that appear in geological time are beyond our powers of conceptualization, but they are accepted by the scientific community as accurate. Reading articles on the subject, one notices discrepancies of thousands of years from author to author, but these discrepancies are infinitesimal when viewed in the light of the vast stretches of time that we are dealing with.

"Eons" are units of time 500 million years in length. "Eras" are units several hundred million years in length. "Epochs" are units of tens of millions of years. Below epochs are "Ages".

The age of the earth was once a matter of controversy, but, as has been mentioned already, radiometric dating indicates that the earth began about 4,570 million years ago. This date marks the beginning of what is known as the Precambrian Eon. There is little fossil evidence to tell us about what was going on at this time, but at about 3,800 million years ago there is evidence that life had appeared in this void in the form of some sort of bacteria.

The lands of the earth were somehow huddled together in a sort of monster continent called **Pangaea**. The modern method to gauge time periods is to classify them according to the fossil evidence found. No matter how distant or how different in material composition, if strata contain similar fossils, it can be presumed that they were laid down at the same time when the earth was just one land mass.

ERA	PERIOD	EPOCH	MYA	MY
PALEOZOIC	Cambrian		544	34
Gk.: Paleos = old	Ordovician		510	71
Zoe = Life	Silurian		439	30
	Devonian		409	46
	Carboniferous		363	73
	Permian		290	40
MESOZOIC	Triassic		250	48
Gk.:Meso = Middle	Jurassic		202	61
	Cretaceous		141	76
CENOZOIC	Tertiary	Paleocene	65	10
Gk.: Kaino = recent	(The Paleogene)	Eocene	55	21
		Oligocene	34	11
		Miocene	23	18
		Pliocene	5	3
Man appears	Quaternary	Pleistocene	2	2
	(Neogene)	Holocene	The recent 10,000 yrs.	

The Pre-Cambrian Eon thus lasted for 3,956 MY. After this "eon," time was counted in eras, periods and epochs. **Keep in mind that our A.D. history has been going on for only 2,000 years so far.**

The vast expanses of time tabulated above need to be appreciated for a better understanding of evolution. If we were to count to a million, devoting one second to pronouncing each figure, it would take us 11.5 twenty-four- hour days to complete the task. If, instead of one second separating each figure, we were to substitute a space of one year, imagine how strung-out the counting would be! It would seem to go one forever! All sorts of things could have happened during such an unimaginable space of time. The theory of evolution claims that this is exactly what did happen!

In the above table, the time spans are outlined, but to be able to imagine the environment in which evolution was occurring we can search the web sites and find expert descriptions of what happened in each period. It was a wild world where earthquakes rocked the globe as continents convoluted and separated from or collided with one another. Ice ages of millions of years lowered sea levels hundreds of meters, wiping out wide ranges of creatures, but eventually thawing and opening up new territories where life forms could multiply again. There were impact events, one of which led to the moon's being ripped off the earth and jettisoned up to its present orbit.

The names of the geological time spans are taken mainly from places where archaeological studies were being made. E.g. "Cambria" was the Roman name for Wales. The Ordovichi and the

Siluri were ancient Welsh tribes. The Permian period in which the greatest extinction of life of all time took place was named after Perm, a town in Russia. The Jurassic Period, when dinosaurs walked the earth, was named after the Jura Mountains between France and Switzerland etc. Then come seven epochs whose names end in "cene". This latter word means "recent", so the epochs mark the degrees of recentness starting 65 mya, the point of time when the dinosaurs were forced into extinction, giving way to mammals to increase and multiply.

At first, creatures had evolved in the waters ("the primeval soup") after the 4,000 million-year Precambrian Eon during which bacteria and algae had produced enough oxygen in the atmosphere to support the "Cambrian Explosion" of life forms. Fossils indicate that marine types gradually sought to dwell on earth, where the flora was forming a habitat to welcome them all. Huge forests laid down the deposits of coal and oil that we have been exploiting for several hundred years. At last, mammals became dominant on earth and their evolution has been mapped from 50 mya when the monkey-like primates began to evolve.

Paleontologists have fossil evidence of evolution up to 200,000 years ago when modern man made his appearance. By the time of the Pleistocene epoch (2 mya) most animals we know had appeared. There were big changes in morphology e.g. the original horses were not much bigger than cats!

However, this booklet is not meant to be a course on geology. Suffice it here to point out that the time spans, and terrestrial upheavals in which evolution was occurring were huge, lavish, relentless and almost indescribable. After Pangaea (the immense land mass extending to both hemispheres) began to break up, the land in the Southern Hemisphere centering around Africa gradually receded towards Antarctica.

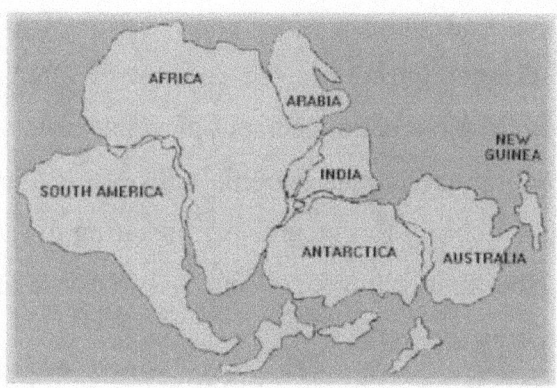

It is called the Gondwana.(Named after a region in India) and is said to have formed 500 million years ago. We can imagine the extent of geological disturbance when India broke off from Africa and Antarctica and was jolted into its present position in Asia, pushing up the land mass in the Himalaya Mountains. Australia was flung to the east across water that became the Indian Ocean, while South America was pushed to the west and stopped an ocean's breadth away, joined up by a narrow isthmus to North America that had remained in the northern hemisphere after the break-up of Pangaea.

Chapter 4

The DNA that Darwin Didn't Know About.

This chapter introduces the concept of DNA. Francis Collins, in his speech at the announcement of the completion of the mapping of the DNA in the human genome, said, " It is a happy day for the world. It is humbling for me and awe-inspiring to realize that we have caught the first glimpse of our own instruction book, previously known only to God."[14]

The author has already introduced the readers to the National Geographic article **Was Darwin Wrong** by David Quammen. The latter implied that "No! Darwin wasn't wrong." However, he did say that, especially as regards modern genetic studies, Darwin was handicapped by lack of information. The Neo-Darwinists now have this information and combine it with classical Darwinism in their research.

For Darwin, natural selection of species endowed with adaptations selected from small, random and hereditary genetic mutations was the <u>only </u>way evolution of life species would have taken place. However, it should be noted that, while random advantageous mutations would have advanced evolution by being naturally selected, it seems that there are also other forces at work contributing to and regulating the evolutionary process.

The discovery that DNA is the central force behind the

---

[14] Cf. *The Language of God,* by Francis Collins p.3

production of biological life forms makes it necessary for readers to have a basic appreciation of what DNA is. It is a molecule called "Deoxyribonucleic Acid" and it contains the biological instructions that make each unique species what it is. In the late 1800s, at the same time as Mendel was working on his plants to show the facts of heredity, a Swiss biologist, **Friedrich Meischer** was studying the chemical make-up of cells. He felt that heredity was somehow bound up with a material that would be found in cells. As part of his research, he took white blood cells and isolated their nuclei to study what protein was inside. However, instead of protein, he found slightly acidic material which he called nuclein. Further study showed that phosphate, sugar and nitrogen were the ingredients of nuclein. He was thus the discoverer of DNA in 1868, but knew little about its purpose.

At first, biologists felt that the DNA molecule was too simple to play a big part in the production of life forms. They felt that this function was the prerogative of proteins, which are far more complex entities. It wasn't until 1953 that the importance of DNA was understood. **James Watson, Francis Crick and Maurice Wilkins** won the Nobel Prize in that year for their discovery of **"the double helix"** structure that enables DNA to carry information from one generation to the next through cell replication.

The existence of DNA was known, but its molecular physiognomy and function were a mystery. Rosalind Franklin was endeavoring to create an X-ray photograph of the DNA molecule, but her studies were still incomplete. Nevertheless, her research gave James Watson some good hints that it might be spiral or helix in shape. He had been a whiz-kid on an American quiz program. He entered the University of Chicago at the age of fifteen, finished his bachelor's degree in biology at the age of nineteen and gained his Ph.D. at the age of twenty-two at the University of Indiana. On the way, he learned a lot about biochemistry and radiation genetics and resolved to get DNA to reveal its secrets. He encountered Maurice Wilkins, the head of a laboratory at King's College, London. After that, he developed a great desire to study at Cambridge, the alma mater of Darwin and where interest in the origin of life was high. Wilkins remained a valued mentor with great faith in the young researcher from America. He cooperated with Watson in every way he could.

At Cambridge, Watson formed a partnership with Francis Crick, a theoretical physicist, whose knowledge of chemistry was minimal. He had used advanced mathematics to gain insights into the structure of proteins by using X-rays. Stephen Meyer, the author of *The Signature in the Cell,* says of Watson and Crick that "they left data gathering to others while they focused on the big picture" of using new thinking to solve the riddles that, for such a long time, had tantalized scientists seeking the origin of life.

Meyer continues, " On April 25, 1953, a seemingly modest paper appeared in the journal, *Nature.* The article was only

900 words long and was signed by two "unknowns" (Watson & Crick) and titled " *Molecular Structure of Nucleic Acids: A Structure for Deoxyribose Nucleic Acid.*" The structure was "a double helix" a shape like a ladder twisted into a cork-screw shape. **This article has revolutionized biology** even more than Darwin's theory had done.

The DNA material is made of elements called nucleotides. It looks like a twisted ladder, the rungs of which are called base pairs because they are formed by pairs of chemicals that combine to form each rung. These chemicals are numbered by the four-letter DNA alphabet, A,T,G,C. The rung elements join together in a specific order. Adenine,"A" always pairs with thymine,"T"; and cytocine, "C" always pairs with guanine,"G." The order in which the rungs are arranged determines what biological instructions are contained in a sequence of DNA (a gene). The nucleotides or the elements in the rungs of the helix ladder are the same in all animals and plants. The order in which they are arranged makes the difference.

The following is a schematic presentation showing a section of a DNA molecule. At the right we see the twisted ladder (Double Helix) each of whose rungs are comprised of two of the chemicals already mentioned. They are joined by light hydrogen bonds. The sides of the ladder are sugar-phosphate. At the left, we see the spiral ladder untwisted and flattened out. The schema is available under " DNA Structure" in Google, along with scores of alternative schemas.

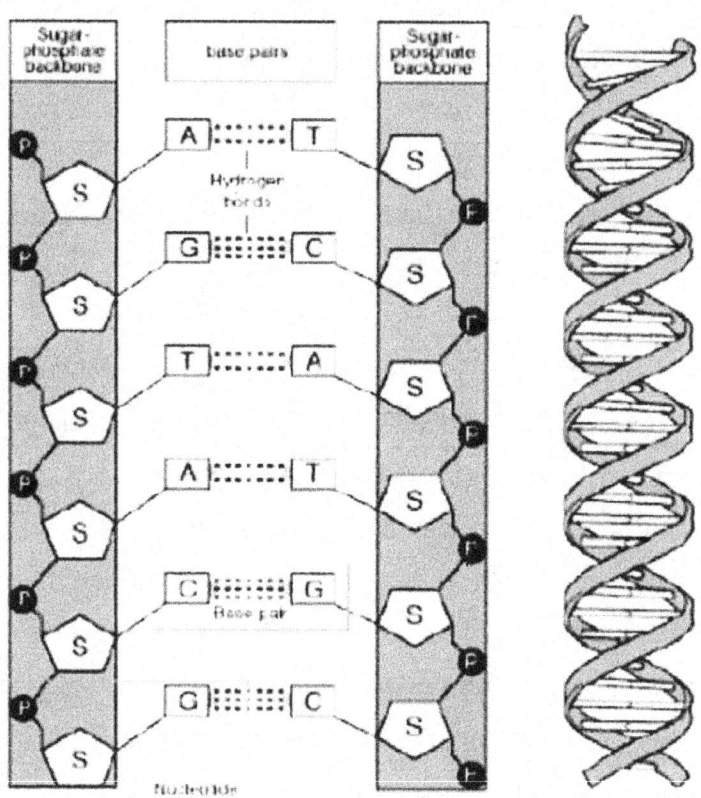

It was Crick, in particular, who surmised that the arrangement of the base pairs in the DNA was probably a code which would further affect the production of proteins in the cells of the organism. **Crick's idea that DNA was an information carrier rather than merely a chemical substance was what made it so revolutionary.** His suspicion of the fact that DNA was a code set off a search for what it coded. Crick felt that cells needed a means of translating and expressing the information putatively stored in DNA, but it took him, with the help of a world-wide spectrum of scientists, five years to verify his suspicion that there must be molecules, intermediate between DNA and the synthesis of proteins, that are regulated by the DNA genetic code.

A two-step process is involved: <u>Transcription</u> of the information contained in the DNA and then its <u>translation</u> into the various end-product types of proteins that contribute to the make up biological organisms.

During transcription, the information contained in a gene (one gene for each protein) is transferred to a similar molecule to that of DNA. It is called Messenger RNA (ribonucleic acid) and is slightly different to DNA in chemical properties. Instead of thymine(T), RNA uses uricil (U).

The process of getting from a gene to a protein takes place in the cytoplasm. A factory-like complex called the ribosome reads the information conveyed by the mRNA and "manufactures" the amino acids which are the building blocks of protein. A type of RNA called transfer RNA (tRNA) assembles the amino acids one at a time to manufacture the desired protein. This process is called the "central dogma" of molecular biology.

Of the protein that is manufactured under the direction of the DNA code, some remains in the cell cytoplasm and the rest is released into the extracellular space under the influence of complex entities in the cell called the Golgi apparatus. From there it is transported to the parts of the body for which it was made.

Meyer introduced a metaphor to make this process more understandable. Somebody sends a doctor a CD disc with step-by-step information on how to treat a patient suffering from a rare disease. The doctor is grateful for the information but he doesn't have a computer at the outpost where his makeshift clinic has been set up, so he can't convert the digitalized information into instructions in the language he knows. In other words, a code is no use unless it works in tandem with a system that can transcribe it and then translate it into the processes for which it was designed to accomplish. In other words, DNA is a code that is useless unless it is complemented by the protein- synthesizing system in the cell.

There are millions of cells in the human body, but, in every cell, there is a complete copy of the map or plan or code according to which the whole body is built. Though nobody knows the number of cells in the human body, it seems that 50 million million (50 trillion) would be no exaggeration. We live in an age when nanotechnology is often referred to. However, we have always lived in an age of super nanotechnology, but we didn't realize it. In *The Selfish Gene,* Richard Dawkins expresses it well, saying that it is as though every room in a huge building (50 trillion rooms) had a bookcase in it with the plans for the whole building are kept. There would be 46 volumes needed to contain the building plans and each page in the volumes could be likened to a gene embracing the building plans for specific, different sections of the building.

The following is a schematic drawing of a cell.

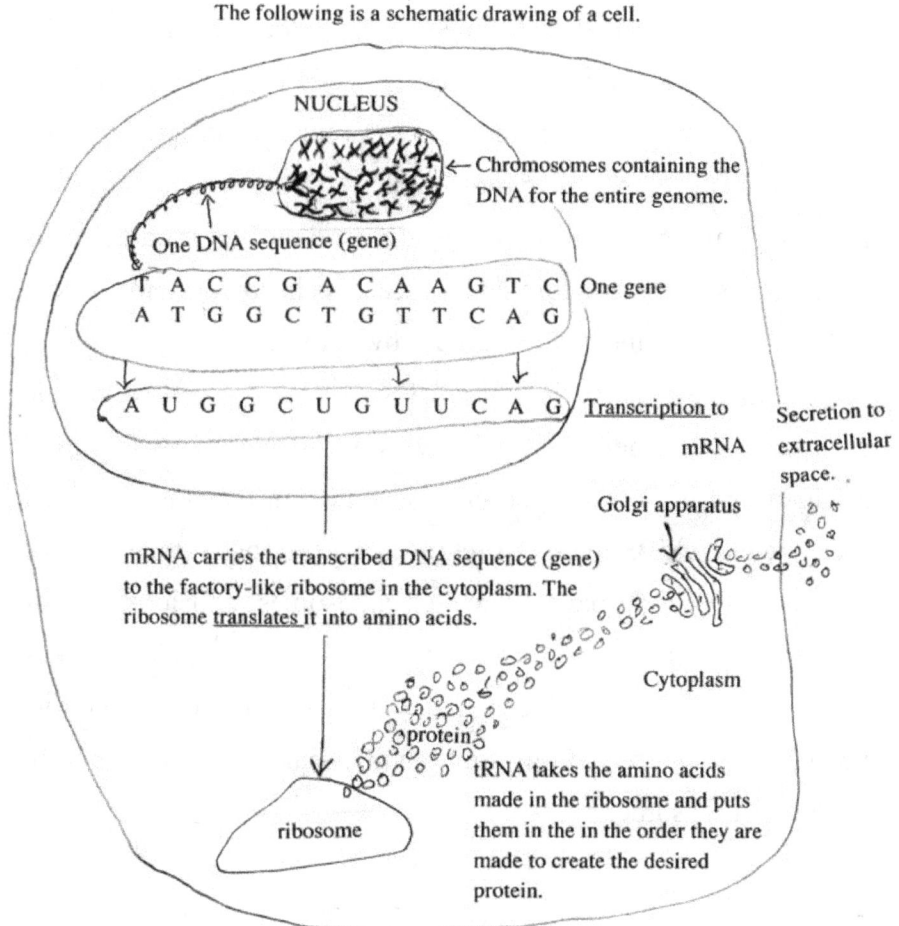

When a baby is first conceived, it is just a single cell incorporating half of its DNA from its father and the other half from its mother (meiosis) and thus receives a DNA code of its own.[15] After conception, there is complex activity in the nucleus of

---

[15] An example of meiosis in which children inherit DNA from both parents would be eye-color. Their eyes will be blue if both parents have

the cell, resulting in the production of two complete sets of DNA (Mitosis). A nuclear envelope forms around each set and, at this time, by a process called "cytokinesis," the extra-nuclear part of the cell is shared out to the two "daughter" cells that will result from the complete division of the cell. The parent cell no longer exists because it has become the two daughter cells. Each cell has a copy of the original DNA. This cell division continues until all the cells in the baby's body are formed and each cell contains the whole DNA code that engineers the whole growth of the baby.

DNA is found in a special area of every cell called the nucleus. Because the cell itself is very small, the DNA is tightly packaged into what are called chromosomes. In humans, each cell usually contains 23 pairs of chromosomes, for a total of 46. Twenty-two of these pairs are called autosomes, and look similar whether they are in males or females. The 23rd pair, the sex chromosomes, differs between males and females. Females have two copies of the "x" chromosome, while males have one "x" and one "y" chromosome.

---

blue eyes, but if the color of the parents' eyes differ, the child will inherit the eye-color of only one of the parents. This is caused by the "dominant" gene and the gene that is temporarily shelved is called the "recessive" gene. This fact is important for medical use of genetics.

Chromosomes

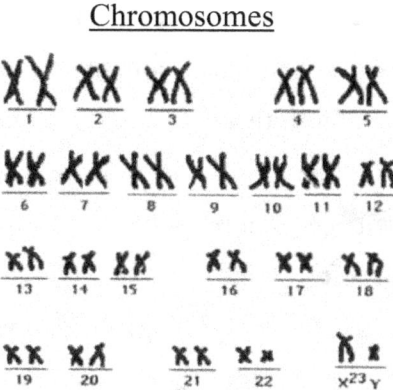

The various sections of the human body have developed in accord with instructions incorporated in certain sections or sequences of nucleotides in the DNA . These sequences of DNA are like sections in an instruction book for building particular parts of the human body. They are called **genes.** Electron microscopy enables us to see the genes occupying definite locations in the DNA strands emerging from the chromosomes, and gene manipulation is one of the goals of modern medicine.

In humans there are about 25,000 genes.

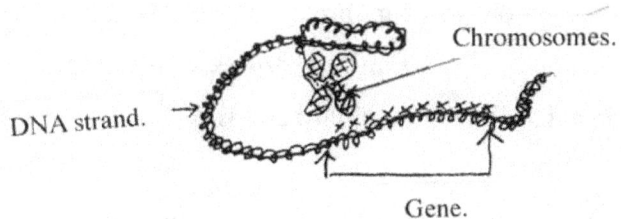

Chromosomes.

DNA strand.

Gene.

In the emerging fields of genomics, scientists have mapped the complete human genome (Completed in 2000 A.D.).

The first Head of the project was James Watson, the co-discoverer of the nature of DNA, but he resigned over the intention of some of the participants to patent their findings. He was replaced by a doctor of similar mind to himself, Francis Collins, who ensured that the findings of the Genome Project would be available to the whole world. He described the genome as a "repository of instructions" and " the book of life". His words[16]on p.123 of his book, *The Language of God,* are worth quoting: " For me as a believer, the uncovering of the human genome had special significance. It is a book written in the DNA language by which God spoke life into being. I felt an overwhelming sense of awe in surveying this most significant of all biological texts. It is written in a language we understand poorly and it will take decades, if not centuries, to understand its instructions, but we have crossed a one-way bridge into profoundly new territory."

Collins points out that a time is coming soon when the genetic glitches that place each of us at risk for some future illness will be discovered, and we may each have an opportunity to find out what is hiding within our own personal DNA instruction book. [17] Personal DNA investigation for e.g. breast cancer, is already being used by a lot of women whose family histories of the disease gives them reason to fear that they also might be at risk.

The genome map would basically look something like the following; TACGATCGGCATATTAGCCG etc going on almost to

---

[16] P.123, *The Language of God* by Francis Collins.
[17] *The Language of God* p.239

32

infinity. Collins[18] tells us, " the human genome has 3 billion letters. A live reading at the rate of 3 letters per second would take 31 years." Specialists can read it and, by comparing the sequences with their products in the organism, they can, hopefully, use their knowledge to manipulate the sequences to cure hereditary diseases etc."[19] Proteins that can act like scissors or glue have enabled scientists to manipulate DNA and RNA by stitching together bits and pieces of these instructional molecules from different sources. This genetic manipulation is called "recombinant DNA." This is a new science, ushering in a new era of civilization. The digital code we use for TV, music, communications etc is made up of ones and zeros e.g. 1001001010001100etc. It is called the " binary code" and uses only two letters, so it is simpler than the four-letter code of the DNA. [20]

There are some very interesting visualizations of DNA that can be found on Google under: DNA Structure, DNA Replication, Mitosis, Molecular Visualization of DNA or "How Big is Your Genome? etc.

At this point we must return to the question raised at the start of this chapter viz. Was Darwin right when he claimed that **evolution is driven only by natural selection of <u>random</u> and heritable genetic changes?** The answer is "No, there are also other forces regulating   evolution.

---

[18] Author of *The Language of God* p.1 of the introduction.
[19] *The Language of God* by Francis Collins p.106
[20] Google. A good explanation of genomics.   Barry Schuler

Chapter 5.

Other Forces Regulating Evolution

When DNA replicates in the course of its usual activity, the possibility of a copying error is not hard to imagine. Darwin didn't know anything about DNA, so he used the terminology at the time, saying that a beneficial genetic change would be made permanent by natural selection and, gradually, a better species would evolve. Francis Collins states, "Darwin didn't know the mechanism of evolution, but we can now see that the variations he postulated are supported by naturally occurring mutations in DNA." [21] The next two chapters are not easy for amateurs to understand. However, what they are saying is that Darwin's "selection of advantageous mutations theory" is not the only source of evolution. Only a small part of DNA is required for the production of proteins, the building blocks of biological bodies. Other regulating factors have also been discovered to contribute to biological evolution. That is the concept that needs to be accepted by reading the next two chapters. 100% comprehension is not necessary.

Collins explains that an overall view of the human genome shows that mutations have arisen at a rate of about one mutation every 100 million base pairs per generation. One in 100 million base pairs seems infinitesimal, but it must be remembered that humans

---

[21] *The Language of God* p.131

have two genomes (one from each parent) and that each genome contains 3 billion letters representing Thymine, Adenine, Cytosine and Guanine, totaling 6 billion letters in all. Divide 6 billion by 100 million and we get 60. This means that in each generation each of us has 60 mutations that are not present in either of our parents.

Some of these mutations are harmful and account for certain genetic diseases. Collins recounts his experience in discovering the mutation that causes Cystic Fibrosis(CF) in Europe.[22]

Many mutations affect parts of non-protein-coding areas in the DNA, and thus cause no basic genetic damage.

Sometimes, a mutation that leads to an advantageous change is selected and passed on to many offspring. This finally results in a degree of evolution that eventually brings a new species into being.

Nevertheless, there is evidence that the concept of Darwinian evolution as stated above needs to be complemented by research results concerning other dynamic forces in DNA.

Despite Darwin's theory of gradual evolutionary progress due to a rather regular pattern (60 per one generation) of random

---

[22] First his team found that it was due to a gene of the recessive pattern (reducing by 50% the chances that it would occur if only one parent had the defective gene.). Then they found that the defect was in a gene in Chromosome 7 where three letters of the code (CTT) had been deleted. Collins himself was one of the doctors who discovered this error. It had taken doctors all over the world ten years to find this mistake in the DNA. Once found, the knowledge was there for possible medical intervention.

genetic changes, and the natural selection of advantageous ones, it seems that paleontologists such as **Stephen Jay Gould** (studying land snails) (1941-2002.)[23] had found a discrepancy between what would be expected from the Darwinian theory in the fossil record and what the fossil record actually presented.

Gould was an American paleontologist, evolutionary biologist and historian of science. He was also one of the most influential and widely read writers of popular science of his generation. He spent most of his career teaching at Harvard University and working at the American Museum of Natural History in New York.

His greatest contribution to science was the theory of **Punctuated Equilibrium** which he developed with Niles Eldridge[24] (who was studying sea or lake floor-dwelling invertebrates, trilobites.) in 1972. The theory proposes that most evolution is marked by long periods (6 million years in the case of trilobites) of evolutionary stability, which is punctuated by rare instances of branching evolution (Cladogenesis). It seems that during these long years of stability, random mutations either didn't occur or, if they occurred, they were not selected. The theory was contrasted against the popular idea that evolutionary change is marked by a pattern of smooth and continuous change. **(Gould, claims in the April 23 issue of *Science*, 2002, that "neither of Darwinism's two central themes, random genetic changes and natural selection, will survive in their strict formulation.** [25])

Now, another of Darwin's ideas also seems to be under fire. Darwin held that the only pathway to evolution was random

---

[23] Google. " Stephen J.Gould"
[24] Cf.Ian Tattersall p. 91-95 *Becoming Human*
[25] Cf. P.14

genetic change made permanent by natural selection. Almost contemporaneous with Darwin was a French zoologist named Jean-Baptiste Lamarck (1744 – 1829)[26]. He is known as the originator of a discredited theory of hereditary, **the inheritance of acquired traits**.

Darwin admired him, even though he didn't accept Lamarck's theory. Darwin wrote of him; " He published his views in 1801. He aroused interest in the view that changes in the organic world were the result of law, and not of miraculous interposition."

Lamarck believed that organisms are not passively altered by their environment. Instead, a change in the environment causes changes in the needs of organisms living in that environment, so they have to adapt themselves and change. He held that such adaptations were heritable.   In other words, the physiological needs of organisms, created by their interactions with the environment are what drives Lamarckian evolution. There would be no time for waiting around until random mutations made it possible for the organism to select the changes it might need for surviving in the hostile environment .

Lamarck's ideas have spent a long period incubating and gathering dust on the shelves of the zoology sections in science libraries, but suddenly his ideas are evoking quite a lot of new interest[27] (e.g. Sharon Begley's) The subject of Begley's article is the lowly "water flea"   "Some of the latter sport a spiny helmet that deters predators; others, with identical DNA sequences, have bare

---

[26] Google " Jean-Baptiste Lamarck (1744 – 1829)
[27] Google, " The Sins of the Fathers, Take 2" by Sharon Begley, 2009.

heads. The difference between the two types are not their original genes but their genes as influenced by their mothers' experiences. If the mother lived her life unthreatened, her offspring have no helmets, but dangerous experience by the mother has led her to adapt to the harmful environmental change. This adaptation has affected her DNA which, in turn, is transmitted to her offspring. "

The article goes on to introduce information about experimental mice studies in Queensland that reach the same conclusion. If mother mouse eats a diet rich in vitamin B12, folic acid or genistein (found in soy), her offspring are slim, healthy and brown—even though they carry a gene that makes them fat, at risk of diabetes and cancer, and yellow. It turns out that the vitamins slap a molecular "off" switch on the obesity/diabetes/yellow-fur gene. (Don't try this at home: no one knows which human genes soy, B12 and folic acid might silence.) This was the first evidence, now confirmed multiple times, that an experience of the mother (in this case, what she eats) can reach into the DNA of her eggs and alter the genes her pups inherit. "There can be a molecular memory of the parent's experience, in this case, diet," says Emma Whitelaw of Queensland Institute of Medical Research, who did the first of these mouse studies. "It fits with Lamarck because it's the inheritance of a trait the parent acquired."[28] Surely, Lamarck is smiling in his grave and saying that it is about time people have recognized his hypothesis that genetic inheritance of acquired traits can indeed help to regulate evolution. **"The environment and the way organisms**

---

[28] From the same source as the above.

adapt to it sometimes affect the genome." Experiments show that changes in the DNA brought about in this way are also passed on to further generations.

This information shows that there are forces involved in evolution that rule out the need for postulating that selection of random genetic change is the sole path to evolution.

Another scientist, **Barbara McClintock (1902–1992)** worked on heredity in the days before the genetic code in the double helix was discovered in 1953. Her work was carried out by experiment, close observation and careful recording of hereditary changes in maize. She was educated and taught at Cornell University. Later, she taught at the University of Missouri and then spent the rest of her life doing research at the Carnegie Institute of Washington's Department of Genetics.

While observing and experimenting with variations in the coloration of maize kernels, she observed the large chromosomes of corn under the microscope. In the 1940's, she discovered that if she damaged the DNA in corn maize, the plant would reconstruct the damaged section by making copies of other parts of the DNA strand, then pasting them into the damaged area.

While experimenting in this way, she discovered two genes that she called "controlling elements". These elements controlled the genes that were responsible for pigmentation.

These controlling elements could move along the gene sequences in the chromosomes, so she called them "transposable genes" or transposons. She regarded them as regulating the genes involved in pigmentation. In 1983, she received the Nobel Prize for her discovery of "mobile genetic elements." It was thought that these elements would be shown to play a big role in regulating evolution, but, by 1970, they had become known as "genomic parasites" participating in the qualities of viruses. Under certain stresses experienced by an organism, they would insert themselves in slots opened up by genes. Somebody explained their activity as tantamount to the plans for the seats of an aeroplane being substituted for the plans to be used for making the engines. The transposons could induce harmful mutations leading to disease.

Since McClintiock's time, a lot of research has been carried out on mobile elements, and much information about the dynamic nature of DNA regulatory function has been accumulated. Before McClintock's time, the accepted view was that genes were static like beads on a string, but McClintock's research, though originally misunderstood even by herself, opened up a whole range of evidence that proved that it was not the case.

Apropos of this, there is an article of interest [29] for anybody interested in the subject of this chapter. James A Shapiro is a Professor in the Department of Biochemistry and Molecular Biology at the University of Chicago. He is interested in evolution by natural genetic engineering.

He says the possibility of a non-Darwinian scientific theory of evolution is something that is almost never discussed. He also introduces the concept of a convergence between biology

---

[29] Google: "A Third Way" by James A.Shapiro

and information science. He states that computer mechanisms are similar to molecular interactions within cells and these intra-cell molecular interactivities were not included in the mechanical concepts that pervade classical Darwinism. For example, in the cell there is an array of repair systems that remove accidental mutations. Proof-reading mechanisms remove errors that naturally might occur during the replication of DNA.

He exemplifies this by using the "ciliated protozoan Oxytricha", [30] which, placed under certain stresses, reorganize their genetic apparatus within a single generation, fragmenting the germ-like chromosomes into thousands of pieces and then reassembling a subset of them into a distinct kind of new genome."

He recognizes the studies of Barbara McClintock as the original source of this **new approach which replaces a constant genome which is subject to random localized changes at a fairly constant mutation rate, by a fluid genome, subject to episodic, sometimes massive, DNA regulated genetic changes.** He points out that our views on the evolutionary process cannot but be deeply influenced by this change of understanding.

Then he points out that there is a growing realization that **cells have molecular "computing networks"** which process information about internal operations and the external environment to make 'decisions' concerning, growth, movement and differentiation.

---

[30]   Google: "Genomes. Oxytricha trifallax. The Genome Center at Washington." (Under the paragraph heading, "Biology")

In another paper aimed more at professional readers[31] Professor Shapiro states: "The principles underlying cellular computing may well be different from those that operate in electronic digital computers, but such a difference doesn't invalidate the information metaphor. However, it means that we have to be careful in applying existing computational models to cells.......... No modern man-made contrivance operates with either the degree of complexity, the precision, or the efficiency of living cells."

Dr Shapiro seems to maintain that there is more evolutionary potential in the dynamic nature of DNA as first hinted at by McClintock's discoveries about the dynamism in DNA than in the random-mutation ideas of Darwinism. The latter place limitations on the pace of evolution, the former allow for freedom to make adaptations at any time the organism requires them for survival in the environment in which they exist.

Junk DNA

The protein coding DNA involved in the manufacture of proteins occupies less than 3% of the DNA in any of our cells. The rest seem to have little or no significance. This led to the slur that called it Junk DNA. These sequences of non-coding DNA have remained the same for millions of years, so they must be doing something, even though biologists haven't been able to pin-point all of what it is that they do. (Non-coding RNA (ncRNA) is RNA that is not translated into a protein.) Cf.p.407 of *Signature in the Cell* by

---

[31]   Google : "A 21st Century View of Evolution: Genome System Architecture"

Stephen Meyer.[32]

James Noonan at Yale University[33] says, " Sections of these non-protein-coding DNA regions in the genome are now known to play a regulatory role, dialing down or cranking up the activity of actual genes."…….. "Researchers identified a DNA region made up of 546 base pairs, or "letters" of code, which have barely changed during the evolution of vertebrates. However, that region has accumulated 16 changes since the ancestors of chimps and humans diverged about 6 million years ago. Thirteen of these changes were clustered within an 81-base-pair region." In other words the genes were stimulated to an episodic degree of activity at the time chimps and humans diverged, probably under the influence of DNA sequences that once would have been identified as junk.

The work of the Queensland geneticist, **John Mattick,** (Institute for Molecular Bioscience. University of Queensland) backs up these ideas: "We are exploring the thesis that the genetic programming of higher organisms has been fundamentally misunderstood for the past 50 years, because of the assumption that most genetic information is transacted by proteins. It is known that only a tiny fraction of the human genome encodes proteins, and the number and repertoire of protein-coding genes has remained largely

---

[32] He points out that the so-called junk DNA 1) Regulates DNA replication.2) Regulates transcription(copy) of the DNA code by RNA polymerase into the RNA format. 3) Influences the proper folding of chromosomes.4) Controls RNA processing, editing, splicing.5) Modulates translation of the RNA transcript into the construction of protein. 6) Regulates embryological development.7) Repairs DNA. 8) Helps in immuno-defense or fighting disease. Etc.etc.

[33] Google: Cf. Junk DNA helps distinguish people from other primates.

static from simple worms to humans, despite huge differences in developmental and cognitive complexity.

On the other hand, the extent of non-protein-coding DNA sequences, traditionally thought of as being mainly junk, has been found to have increased with great complexity. Moreover it is now evident that these sequences are transcribed in <u>a dynamic</u> manner, and that most complex genetic phenomena are ncRNA-directed, which suggests that there exists a vast hidden layer of regulatory RNAs that control human development."[34] This is expressed in easier words by Perry Marshall : "Scientists are now speculating that proteins and regular DNA that creates them, are just the nuts and bolts of the system. They're like the parts for a 757 jet sitting on the floor of the factory. The non-coding DNA is likely the assembly plans and control systems." [35]

We have tried to produce in this chapter a case for the idea that Darwin's "random mutations" are not the only ingredients involved in evolution.

We have also tentatively introduced a concept that DNA activity can be compared to that of computer technology. The next chapter will take a closer look at this metaphor.

---

[34] Google: Rnomics: RNA in Mammalian Evolution. Cf.p..407 of *Signature in the Cell.*
[35] Google: Cf. Cosmicfingerprints.com

Chapter 6

Computers for Dummies

Francis Collins says that to plead ignorance of how self-replicating, information-storing molecules came into existence is the most realistic attitude for us to take at present.[36] Nevertheless, he warns that to invoke a supernatural cause or, what he calls, a "God of the Gaps" to explain this phenomenon may be premature. Collins writes, "Science is the only legitimate way to investigate the natural world. Whether probing the structure of the atom, the nature of the cosmos, or the DNA sequences of the human genome, the scientific method is the only way to seek out the truth of natural events..... By nature, science is self-correcting. No major fallacy can long persist in the face of progressive increase in knowledge."[37]

Richard Dawkins (Ch.2 of The Selfish Gene) outlines a theory that atoms produced in the Big Bang linked up to form molecules that perhaps existed even before the advent of the earth. At some point a remarkable molecule that could make copies of itself came into being. These "replicators" are the forerunners of DNA-containing cells.

On the other hand, Stephen Meyer claims, as mentioned later on p.106 ff. that at least the existence of the genetic code requires the existence of an intelligent cause. Following his ideas, it

---

[36] *The Language of God* p.90
[37] Ibidem. P.228

would seem that God stepped into biology when DNA came into being, just as He had stepped in to cause "The Big Bang" that launched the universe. Could we call DNA, the Big Bang of biology? Or, could we say that DNA came into existence as one result of the Big Bang, just as the solar system, so hospitable to life, had come into being along with the Big Bang.

In this book, we don't attempt to solve the problems that still remain in our knowledge of just how evolution occurs. Instead, the book aims to point out that problems do exist that are only recently being explored and debated. At this stage it could be stated that Darwin's ideas describe the basis of evolution, but dynamic factors that have been identified in DNA are regulatory forces without which evolution would not be able to be properly explained.

Maybe it would be better to get onto firmer ground and speak about DNA as it is known today. When I say, "today," I mean it literally because, for example, in a recent newspaper there is an article that illustrates precisely what I want to say in this chapter.[38] The author of the article is speaking of children mentally disabled due to what is called "Fragile X Syndrome." This syndrome is caused by a genetic mutation in chromosome X. (The mutation in question is a random one, bearing out the observations of doctors engaged in genetic diseases, that mutations often bring about bad results.). Gene instructions in this chromosome are repeated several times. When they are repeated 200 or more times, a message seems

---

[38] The Japan Times, 22nd.June, 2010 p.10, in an article by the journalist Cesar Chelala.

to go out, instructing that the gene be shut down. As a result, the protein that is normally produced by that gene is no longer produced, or, if it is produced, it is defective. This is what causes the wide variety of symptoms among those afflicted with the disease.

The protein normally produced through the intervention of that gene ordinarily acts as a sort of coordinator of information among brain synapses (connections between nerve cells), helping to stop or slow-down brain signaling at critical intervals. Regulating the flow of information among brain cells is crucial for the brain's ability to learn and develop normally." etc.etc.

What I am trying to illustrate here is that biology has been mired in the contention that evolution over the eons, eras and epochs has been the product only of random accidents in DNA replication. However, these days, with advances in genetic studies, briefly outlined in the previous chapter, the process of evolution from microbes to man is now viewed as something that can be regulated by dynamic forces in the DNA that seem to have been programmed into it like the information in a computer program, and which respond to chemically induced signals to actively intervene in all the aspects of the life of an organism.

The computer program that challenges people to a chess match comes to mind. In response to a person's moves, the computer summons up countless optional countermoves and selects one. This continues on and on until the opposing player is vanquished or victorious. Such a program is obviously complex and voluminous, but it would be far simpler than anything paralleling it in the DNA of humans.

We know, of course, that deoxyribonucleic acid (DNA) is not itself an intelligent being, so we are forced to assume that its apparent intelligent activity is due to something in it akin to a computer program. The big difference from the mathematical equations used in ordinary computing, is that DNA is faced with far more complex situations than we encounter in our ordinary experience of information technology used in computing.

Thus, it can be said that, within the DNA, in addition to the code governing the production of the protein building blocks of life, there is also an ingenious program that regulates many other factors involved in the organism's evolution.

Let us take another glance at the protozoa, oxytricha. James A.Shapiro exemplifies the ability of DNA to respond to certain stresses by using the "ciliated protozoan Oxytricha", [39] "which reorganize their genetic apparatus within a single generation, fragmenting the germ-like chromosomes into thousands of pieces and then reassembling a subset of them into a distinct kind of new genome."

---

[39]   Google: "Genomes. Oxytricha trifallax. The Genome Center at Washington." (Under the paragraph Heading, "Biology"). Also, Google: " *A Third Way" by James*

Not yet fully understood activities of non-coding RNA[40] regulate the rearrangement of genes to produce useful adaptations for different environments. This is probably carried out in small steps that adjust variable options until the organism is maximally adapted to its environment.

A specialist in information technology gives his views:

There are some interesting papers by Perry Marshall. He is an expert in information technology and puts a strong case for the need to recognize intelligence as an underlying necessity to explain any code, let alone the genetic code that choreographs the life of an organism. He calls information a mysterious reality, saying that it is a message. Messages themselves are not matter, but can be carried by matter e.g. the slots on a roll of a pianola music score, ink on note- paper, or by the physical energy expended by the human voice, or by the electronic energy of e-mail or the internet. However, the message itself is immaterial. It can be stored or copied in many forms, but its meaning remains the same and it adds no weight to the medium that carries it.

Marshall continues: The materialist world-view has no explanation for the existence of information, because all information comes from intelligent beings. He challenges his readers ("Show me one example of information that has not come from a mind.") He maintains that all codes contain the following four layers:1) Alphabet (symbols or ciphers that stand for it.).   2) Grammar

---

[40] A.Shapiro

(rules about how words are organized). 3) <u>Meaning</u> (All codes have meaning).   4) <u>Intent</u> (There is a larger purpose beyond meaning).

Could the code that DNA carries have occurred by random chance? Some would say that the arrangement of "nucleitide bases" could, given millions of years, possibly be explained in this way, but language is not made from the "alphabet up". It is made by "intent down". "When you speak, you create language from the top down. When you listen, you go from the bottom up. An intent always precedes communication. There is no exception, and the idea that the language in the DNA code could have formed from the bottom up is only wild speculation. It is contrary to experience and science(sic.)" [41] According to Stephen C.Meyer,[42] "Francis Crick published his ideas of the "sequence hypothesis": 'the nucleotide bases function like letters in a written language to convey information, depending on their arrangement, for building proteins.' Richard Dawkins[43] stated, "The   code of the genes is uncannily computer-like." Software developer, Bill Gates [44] said, 'DNA is like a computer program, but far, far more advanced than any software ever created.'"

---

[41] Google :*Perry Marshall's Intelligent Design Quick* Guide (P.6 of 10)
[42] *Signature in the Cell.* P.12 Meyer also points out that scientists have been preoccupied with investigation of matter and energy. However, they also have to incorporate "information" into their way of thinking. A gene is specified by the pattern of base pairs in a DNA molecule, but the DNA molecule is the medium, it's not the message. A CD disc doesn't change weight when it receives information, but its significance is increased to the nth degree.
[43] Dawkins. *River Out of Eden.* P.17
[44] Gates, *The Road Ahead.* P188

A note that should be inserted here is the recent publication of **Craig Venter's success in code manipulation**, as a step to utilizing microbes to produce synthetic fuel. "Digitalizing of Biology" [45]

"We and others have been working for the past several years on the ability to go from reading the genetic code to learning how to write it. It is now possible to design in the computer and then chemically make in the laboratory, very large DNA molecules. A few months ago we published a scientific study in the journal Science where we described the ability to take a chromosome from one bacterium and place it into a second bacterial cell. The result was astonishing - the new DNA that we added changed the species completely from the original one into the species defined by the added DNA. You could describe this as the ultimate in identity theft."

The next chapter will take us on a long evolutionary saga through the dark geological ages in Africa and lead us to ourselves. We will watch the journey, convinced that, in addition to random, advantageous adaptations, we have also been helped on our way by dynamic, computer-program-like regulation by DNA, and we know that understanding of this won't be able to be understood just by reading " Computers for Dummies."

---

[45] Google: Craig Venter and the ultimate "identity theft."

Chapter 7

The Good Old Days in Africa

For this booklet to live up to its title, it should start with the first steps in evolution taken by our original ancestral microbe. This is what Darwin held and is what could have actually happened. However, for lack of definite knowledge, we will skip over a few thousand million years of evolution among micro-organisms and start at about 30 mya with monkeys who are closer to us. **From this point onward, we know that besides Darwinian selection of random advantageous genetic mutations,** there will be stages of equilibrium interrupted only after long periods extending for even millions of years (Gould & Eldridge), inheritance of some acquired traits (Lamarck), timely intervention of regulatory mechanisms organized by a whole range of possible influences by not-yet-fully-understood non-coding RNA (McClintock, Mattick, Craig Venter) etc.etc

Fossil evidence shows that monkeys existed about 30mya. Paleantologists, studying prehistory through fossils in geological time, tell us that the monkeys evolved from proto-primates over 30 mya. (The proto-primates were apparently small, squirrel/opossum-like, insect-eating mammals.). The fossil evidence also indicates that monkeys out-competed the proto-primates and replaced them. The Oligocene (starting 34 mya and lasting for 11my) was a time of major geological change, resulting in the cooling of the northern regions and the disappearance of primates there.

Monkeys evolved from proto-primate ancestors, following the rules of evolution; " cladogenesis" – branching off from their parent lineage in order to better survive in the environment they were in. Minor changes followed one another and former types went extinct. Eventually, after the passage of millions of years, a new species would evolve, completely <u>diverging</u> from and different to the parent stock. However, the monkeys were not the terminus of the line. Monkey fossils are common in the Miocene (starting 23 mya and lasting 18 my) with apes evolving from monkeys. Fossils of monkeys are rather rare in the Miocene, but Ape fossils are common.

Hominids

From here on, the word "hominid" is used as an evolutionary term referring to all fossil individuals, starting with the great apes (hominidae) who are thought to be in a direct line to our ancestral origin. Another note defining hominids is that they have no tails. Gorillas and chimpanzees diverged from the original primate line between 10 and 6 mya. Genetic studies show that the last common ancestor of chimpanzees and humans was about 6 mya. Finally, about 3 to 4 mya, another type of primate made its appearance. It was called **Australopithicus** and its fossils reveal that it was closer to human beings than the other primates in that it walked upright. There are quite a few different types of

Australopithecines that emerged during a period of 4 million years. Some of these went extinct while some evolved to higher forms that eventually saw the rise of modern man.

The fact of evolution was gradually accepted widely, and the word " missing link" was often used in common parlance. Evolution within the species was well accepted, in the sense that animals that escape from predators are usually stronger than those that fall prey to them.   Over the years, the descendents of the stronger ones evolve, not into a new species, but into a stronger version of the same species. The idea that we human beings evolved from another species such as monkeys was often the subject of jokes. However, a serious search was being made for links that bind us to ancestors differing in species to ourselves, ancestors between us and the apes.

A few examples might help at this stage: **Raymond Dart** (1893-1989) was an Australian Professor of human anatomy who was working in South Africa. He obtained a fossil skull from a limestone quarry at Taung in 1924. It was the skull of a child that he called " The Taung Child." The foramen magnum, indicated the Taung child to be bipedal, and it was calculated to have existed between 3 and 2 mya

The foramen magnum[46] is the opening in the skull where the spinal cord enters it. It is at the rear of the skull in quadrupeds

---

[46] Google: Evolution of Homo sapiens.

like dogs, a little further forward in knuckle-walking mammals like chimpanzees, but almost at the center of the skull in bipeds like humans. Its teeth also were more like a human's than an ape's. He called it Australopithicus africanus (Southern ape from Africa) .

A good idea of the comparative position of the foramen magnum can be gained from the following illustrations:

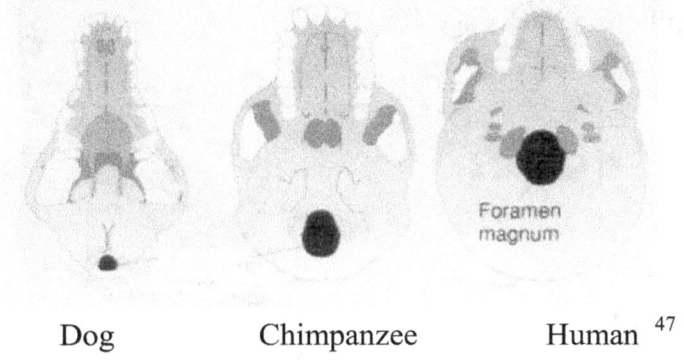

Dog   Chimpanzee   Human [47]

Following Dart was a Scottish doctor working in South Africa. His name was **Robert Broom** and most of his work on fossils was carried on from 1936 through the 1940's. He had been dismissed from a university in 1910 for his insisting on the correctness of Darwin's theory of evolution. In 1936, he discovered the first known adult Australopithicus africanus, in Sterkfontein. He also found fossils of other stronger-looking hominids which scientists now call Australopithicus robustus, but which seems to have evolved outside the line that leads to man.

The above discoveries were made in South Africa, but a

---

[47] Google   Pictures of Foramen Magnum

more fossil-rich area is the Great Rift Valley that runs along the east coast of Africa. The area known as the Olduvai Gorge in the Rift Valley has geological strata easily datable back to 2.1 million years. The remains of many Australopithines and early members of the genus "homo," though differing greatly from our modern human species are plentiful in this region and can be dated by the "potassium-argon" dating method, a viable method to date fossils from millions of years ago.[48]

THE RIFT VALLEY (Shaded)

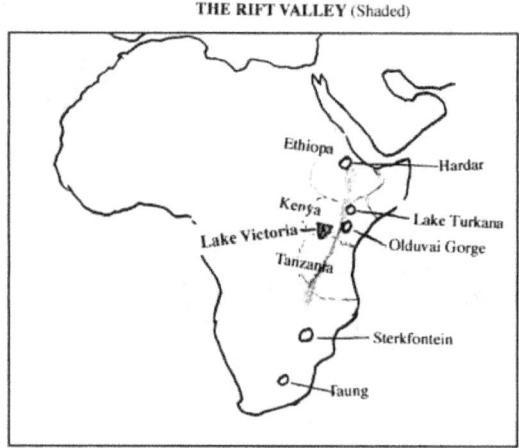

An internet article that makes a very big impact on readers can be found at the following site that treats the discovery of what is probably the most famous hominid fossil, "Lucy". [49] The document tells the story of the discovery of the fossilized Australopithicus aferensis.

---

[48] Google: Chronological Methods – Potassium-Argon Dating.

[49]Google: http://www.becominghuman.org/sites/default/files/englishtrans.pdf

The discoverer's name is **Donald Johanson** (American, and Curator of the Cleveland Museum of Natural History). His words are quoted here. The document recommended gives a good view of how paleontologists work:

"Hadar, in Ethiopia is a place of pilgrimage for those of us who study human origins. It is located in the northern Afar region of the country and can be inhospitable at times when temperatures reach 125 degrees. My first trip to Hadar in 1972 was a short exploratory trip, but I was there long enough to realize that I was going to do field work there. Hadar had the right kind of geology and was very rich in animal fossils dating to about 4 million years ago and I thought to myself that if we could find hominid fossils in deposits this old we might just open up a new chapter shedding light on human evolution.

In 1973 I made my first fossil hominid discovery, a knee joint. The specimen came from a geological stratum dated to nearly 3.4 million years ago. Detailed study of the functional anatomy of the knee showed beyond the shadow of a doubt that it was a creature that walked upright – a hominid. Then in 1974 we returned to Hadar and, in late November, near the end of the field season, I made a discovery that firmly placed Hadar on the map as one of the most significant hominoid fossil sites in the world

I remember very clearly, it was about noon and I had been surveying since just after breakfast. The temperature was approaching 110 degrees. I hadn't found much except a few teeth of a horse, part of a skull of an extinct pig, some antelope molars and a bit of a monkey jaw, but, as I turned to leave a fossil caught my eye, and I knelt down for a closer look.

It was, in fact, part of an elbow. As I looked further I saw another bone and then another. It was truly unbelievable. What I found was a partial skeleton eroding from the ancient Hadar sediments. I knew immediately it was a hominid. A very old one and astonishingly complete.

That night in camp we examined the fossils and in the background the Beatles song; *Lucy in the Sky with Diamonds* played over and over on a small tape recorder. Because of the petite stature of the skeleton I suspected right from the start that it was a female. At some point during the evening the new fossil picked up the nickname Lucy. And she's been known as Lucy ever since.

Until the discovery of Lucy there were few hominid discoveries dating back to more than 3 million years. So Lucy, at 3.2 million years, now offered many new insights to our ancestral past.

Because of her unique anatomy we dubbed the Hadar fossils a new species; ***Autralopithecus afarensis***. Australopithecus means "southern ape" and afarensis celebrates the Afar people in the region where Lucy was discovered.

So much for the words of the discoverer of one of the very important fossils linking modern man to the apes. The discovery has been interpreted with not much disagreement by all the leading paleontologists. From Lucy the search has gone on to the various species that came before and after her in time and who are claimed to have been in the line of our direct ancestors. (In a recent newspaper, the Japan Times June, 23rd, 2010 we read: " Scientists may have found the great, great grandfather of Lucy. A new skeleton has been found in the Afar region of Ethiopia. Dated at 3.6 mya, the find is about 400,000 years older than the famous Lucy. The find was reported in the Proceedings of the National Academy of Sciences on the 22nd June,2010. The bones indicate that the ancestor walked upright but was considerably larger than Lucy, who was only about 1 meter tall). She herself, seems to have followed a species called Ardipithecus who, in turn, followed the gorillas and chimpanzees that branched off the parent stock from about 6 to 10 mya.

Once it was thought that Homo erectus and the Neanderthals were the only two species intervening between us and the primates, but now the population of intermediate qualifiers has grown immensely. There is such an accumulation of fossils that it is hard to keep up with them. Colin Groves (Professor of Bio-anthropology at the Australian National University), uses the following words that are worth quoting at this stage of our investigations.[50]

"It's an exciting time to be alive if you're interested in human evolution. New countries are getting onto the palaeo-anthropological map: India, Syria, Eritrea, Chad, Malawi, Portugal. Every new fossil fulfils certain expectations but opens up a

---

[50] In the conclusion published in the May/June 1999 issue of Reports of the National Center for Science Education,

whole barrel of new research questions. Fossil discoveries are matched by new discoveries of just how human our nearest living relatives are and the press is avid for them all. Keep on your (bipedal) toes; if you miss this week's reports you might already be out-of-date". Paleontology is an exacting science. The following paragraph is meant to illustrate the way paleontological discoveries can help us to understand evolution.

The Japan Times Sept.10th, 2009, reports the discovery in China of a chipmunk-size mammal fossil. The mammal lived 123,000,000 years ago. The three bones of the reptilian middle ear are joined to the jaw hinge, but the bones of this mammal are separated from the jaw hinge. This detail could be important for understanding the evolution of mammalian hearing, an adaptation that helped it survive the dinosaur-infested Mesozoic period (250 – 66 mya). We probably inherited this adaptation. The example also illustrates the minute steps that mark evolution.

The following presents a time-line listing most of the species whose fossils show they existed between the apes and the members of the "Homo genus." To put these species in a time line is relatively easy for experts, because the deposits and strata in which they have been found can be dated fairly accurately. However, to confidently claim knowledge of the relationship between the various species in the time-line is another matter. Depending on which expert you consult, the interpretation of the mutual relationships of species to one another tends to differ.

Some of the mammals in the time line probably had no part in the evolution of man. They arose and then disappeared. The following table, based on paleontological dating methods, includes an approximate estimation of how long ago each Australopithecine type existed on the earth.

The Australopithecines evolved from apes and seem to have evolved into bipedal mammals because climate changes thinned out the lush forests where monkeys thrived. There was more open land between the trees and this engaged the apes in more walking. Natural selection favored genetic changes that made walking with an upright stance easier. However, the Australopithecines still had a long way(several million years) to go before mammals evolved from them that could be categorized into the "Homo" genus.

Their brain size extended between 350 and 450 cubic centimeters. We could list some of them time-wise as follows:

1) 6-7 mya. **Sahelanthropus tchadensis** (Sahara, Chad)Oldest known hominid species.

2) 4.4 - 5.8 mya.   **Ardipithicus ramidus** (Pre-Australopithecine). Fossils are hard to decipher. "Ramidus" means "root."

3) 4.2 – 3.9 mya.   **Australopithicus anamensis** ("anam" = a lake)

4) 3.9 – 3 mya.   **Australopithicus afarensis** ("Afar" people in

N.Ethiopia) (Lucy) .Brain size 375-550cc.

5) 3.5 mya.   **Kenyanthropus platyops** (Flatface from Kenya)

6) 3-2 mya.   **Australopithicus africanus** (Southern ape from Africa) cf.story of Taung Child.

7) 2.5 mya.   **Australopithicus garhi** ("surprise" unusual features) Its large teeth were "surprising".

8) 2.6 – 2.3 mya. **Australopithicus ethiopicus**(Paranthropus=like human) (Black skull stained by manganese).

9) 2 – 1.5 mya.   **Australopithicus** (Paranthropus = like man) **robustus**

10) 2.1 – 1.1 mya. **Australopithicus(Paranthropus)boisei**(strong teeth) "The hominid with nutcracker jaws."   After many millennia Paranthropus disappears from the fossil record. etc

The above species are regarded as the first missing links between chimpanzees and ourselves. The "Homo" genus peaks in Homo sapiens, but the process took about two million years and included many species that the average man today would not regard as "human" in the ordinary sense of the word.

Chapter 8

Still in Africa, the "Homo" Genus begins

The following table presents an abbreviated schema of the evolution of the Homo species, starting with Homo habilis, two million years ago and leading to the rise of Homo sapiens or modern man, 200,000 years ago. The Homo is a genus of primate, the only living species being Homo sapiens or modern man. One of the characteristics of Homo is absence of animal-like body hair.

Species	M.Y.A.	Where	Height	Brain	Fossils
H.habilis	2.2 –1.6	Africa	1..2 m	660 cc	Many
H.erectus	1.4 –0.2	Africa, Java, China, Eurasia.	1.8 m	850 –1100	Many
H.ergaster	1.9 – 1.4	E &S Africa	1.9 m	700 -850	Many
H.antecessor	1.2 – 0.8	Spain	1.75 m	1000	2 sites
H.heidelbergensis	600,000 – 35.000	Europe, Africa.	1.8 m	1100-1400	Many
H.neanderthalensis	250,000– 30,000ya.	Europe, W.Asia	1.6 m	1200-1700	Many
H.sapiens	From 200,000	Worldwide	1.6 m	1000-1850	Still living

1) **2.2– 1.6 mya. Homo habilis** (Handyman) is the earliest known

species of the genus *Homo*. In its appearance and morphology, *H. habilis* is thus the least similar to modern humans of all species in the genus. *H.habilis* was short and had disproportionately long arms compared to modern humans, but it had a less protruding face than the Australopithecines from which it is thought to have evolved.

H. habilis had a brain size (500 – 700 cc.) slightly less than half of the size of modern humans. Despite the ape-like morphology of their bodies, *H. habilis* remains are often accompanied by primitive stone tools. *Habilis* being placed as a species in the Homo genus is because of the primitive intelligence it shows by tool making. Jared Diamond remarks, " The surviving stone tools from the period can only charitably be described as very crude."[51]

3) mya. Brain size: 900-1000   **Homo ergaster**

(the working man) Hard to prove that Ergaster is not just early Erectus Cf. Turkana Boy- the most complete skeleton ever discovered. Ergaster had a rounded cranium, higher forehead and smaller teeth than the Australopithines.

The schema on the next page was a hand-out made by Bruce MacEvoy. It is based on *From Lucy to Language* by Donald Johanson and Blake Edgar.

A close look at the schema shows that some of the Australopithecines seem to have made no input into the development of M3)   **1.4 – 0.2 mya**. Brain size 850 - 1,100 cc. **Homo erectus.**

---

[51] Cf. *The Rise and Fall of the Third Chimpanzee* p.33

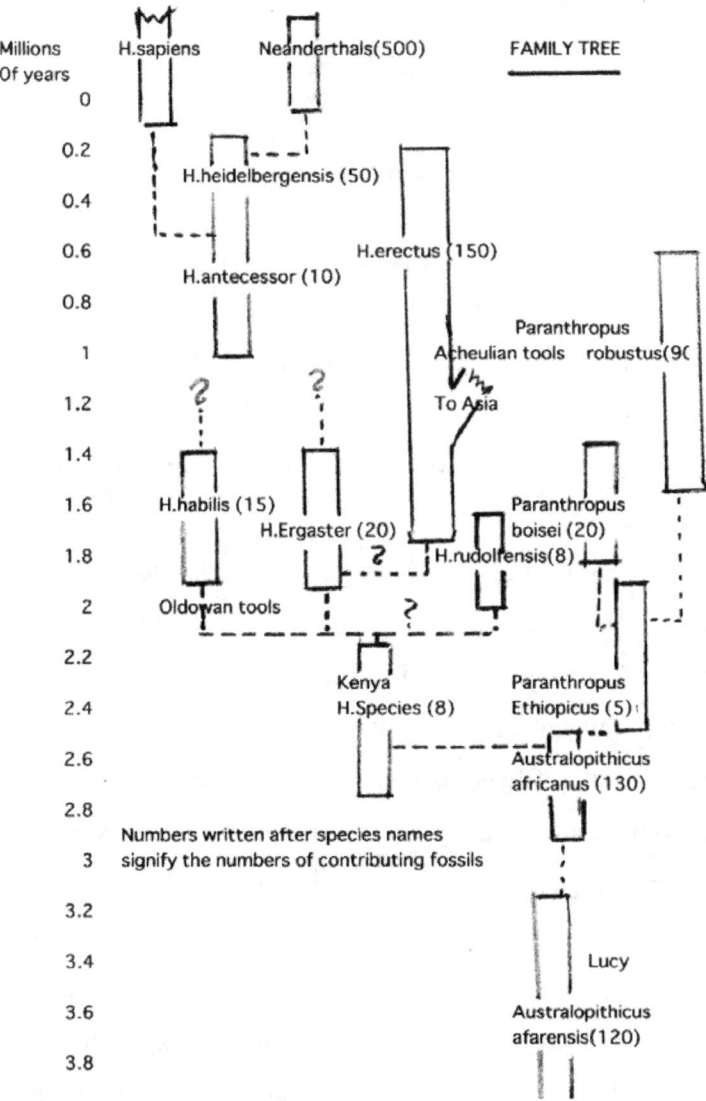

(Standing-upright man) This species is well known and many different species may be subsumed under its name.

Homo erectus skull of Java Man. At Ann Arbor Museum of Natural History. The skull was discovered in Java, in 1891, by Eugene Dubois. The high ridges above the eyes distinguish it from Homo sapiens.

Homo erectus groups migrated out of Africa to Georgia, Indonesia, Vietnam and China. In Georgia, two 1.7my old craniums have been excavated. Outside Africa, Erectus used pre-Acheulian tools, which indicates a very early migration.

The first fossils of Erectus were found not in Africa, but in Indonesia (Java Man) along with primitive Oldowan tools, such as those used by Homo habilis. This does not mean that there was no Homo erectus population left in Africa in the sense that every individual in the species went on the package tour to the Far East. Good fossils were also found near Peking (Peking Man). Most were lost in World War II, but good plaster models had been made. Later, Erectus fossils were found abundantly in East Africa.

Erectus became extinct about 500,000 years ago except in Indonesia where Homo floresiensis (probably evolved from Erectus) survived until 12,000 years ago. Floresiensis would thus have been contemporaneous with Homo sapiens. Erectus had thus wandered the world for over 1.5 million years, finding ample time to multiply and spread

5) 1.2 –0.8 mya. Brain size:1000 cc. **Homo antecessor** (The pioneer leading to humans) Recognized as the oldest European hominid. Previous to the discovery of this species from an accurately dated site 780,000 years ago, the oldest European fossil was that called **Homo Heidelbergensis,** found near Heidelberg and which dated back to 500,000 years ago. However, work on a railway cutting in North West Spain called Gran Dolina, in the Arapuerca region, unearthed a large find of fossils in 1994 comprising six individuals**.**

Some doubted these were from a species separate from Heidelbergensis but, the Spanish paleontologists claim that[52]"Our opinion is that Homo Heidelbergensis is a <u>late </u>Pleistocene European species, which evolved into Neanderthals and not to modern humans. The newly named species *Homo antecessor* (of the <u>Early</u> Pleistocene age; 2.5 million to 12,000 years ago.) is ancestral to *Homo sapiens,* but not directly ancestral to Neanderthals, who arose from Heidelbergensis. Neverthetheless, many paleontologists, regarding Heidelbergensis and Antecessor as of the same species, still often say that Heidelbergensis is the direct ancestor of Homo sapiens!!!

It is thought that Heidelbergensis evolved from and is a

---

[52] Google: http://www.archaeology.org/online/news/gran.dolina.html

migratory descendent of Ergaster. Fossil evidence suggests that its hearing was well developed and that primitive vocalization was probable.

6)   250,000 – 30,000 ya. Brain size 1400 – 1500   **Homo Neanderthalensis**: It is said that paleontology began as a scientific discipline in August, 1856. This was when the specimen named "Neanderthal 1" was discovered in the Neander Valley near Dusseldorf in Germany. This discovery was also the beginning of a long debate that is still going on today. The range of Neanderthal traits is very wide, but it is said that the morphology of their inner ear is unique and serves as an identifier of the species.

Specimens are found over a wide field in Europe and Western Asia, but never in Africa. The ones found in Croatia in 1899 are the largest group of fossils of Neanderthals ever discovered. There are more than 850 fossils there from more than 80 individuals. This material has been dated to 130,000 years ago. Material found in Israel has been dated by thermo-luminescence to approximately 45,000 y.a., making it one of the late survival sites of the species.

Starting in 1997, scientists began investigating the genetic materials found in Neanderthals. Svante Paabo of the Max Planck Institute for Evolutionary Anthropology in Leipzig, Germany, was able to prove that, although the nuclear DNA of Neanderthals and Humans was 99.5% identical, Neanderthals were distant cousins rather than ancestors of humans. They had been different species since their last common ancestor, 500,000 years previously. These studies were based on mtDNA (mitochondrial DNA). The 0.5% difference sounds very small, but in genetics, this variation in "nuclear DNA" can account for a great deal of difference. The Neanderthal arms and legs were much thicker than ours, suggesting immense strength needed to survive the rugged conditions of their lifestyle. However, their bodies were very human in appearance, but their prominent noses, long faces, sloping foreheads and big skulls suggest that they belong to a species that differs from our own. Their big bodies resulted in their large brain capacity (surpassing that of Homo sapiens). However, after coming into contact with Homo sapiens for about 10,000 years starting 40,000 years ago, they went extinct about 30,000 years ago.

The Neanderthals lived in Ice-Age conditions and they hunted Pleistocene animals such as mammoths, competing with larger predators including wolves and lions. The tools they used were effective, but fairly simple, mainly Acheulian, and they changed very little over the period of their existence. This contrasts with the rapid advance of tool-making technology made by early modern humans. However, there is some evidence that they cared for their weaker kin (fossil evidence of individuals suffering crippling disease or injury) and that they buried their dead (in a fetal position). On the Web there are many articles claiming that there was a certain amount of intermarriage between Humans and Neanderthals, but the degree to which this has affected the genome seems to be next to nil.

7)   **160,000 years ago**. Brain size 1,350 cc.:   **Modern man. Homo Sapiens.** Genetic studies and fossil evidence indicate that modern humans **originated in Africa about 200,000 years ago**.   In 1997, Professor Tim White from the University of California found stone tools and an adult skull near a village in Ethiopia called Herto. The fossils were the oldest fossils of modern man (Homo sapiens) yet found. Highly accurate dating showed that the fossils were 160,000 years old; 30,000 years older than the ones previously discovered.   A 195,000- year- old fossil from the Omo 1 site in Ethiopia also shows the rounded skull of the species we usually call H.sapiens. A team from Arizona University found evidence that 164,000 years ago, in a cave in South Africa at Pinnacle point, that there were beings that were eating shell-fish, making complex tools and using red ochre pigments – all modern behaviors. 200,000 years ago, modern man was in Africa.

Chapter 9

Man's 120,000 Year Detour via Asia to Europe

Paleontology tells us that Homo sapiens appeared in Africa approximately 200,000 years ago and entered Europe via the Levant about 40,000 years ago.

For an enthralling account and <u>animated map</u> of the round-about tour taken by Homo sapiens from Africa to Europe, the reader is urged to open the internet on Google at < JOURNEY OF MANKIND – The Peopling of the world.> by Stephen Oppenheimer (**Stephen Oppenheimer** is a world-recognized expert in the synthesis of DNA studies . Publications - Out of Eden, The Real Eve, The Real Eve Discovery Channel DVD, Out of Africa's Eden, Eden in the East.)

Oppenheimer points out that 160,000 years ago, Homo sapiens lived in Africa and the earliest known archaeological evidence, mtDNA and Y chromosome ancestors are found in East Africa.

160,000 - 135,000 years ago groups traveled as hunter gatherers south to the Cape of Good Hope, South West to the Congo Basin, West to the Ivory Coast. They carried the first generation of mtDNA gene type "L1".

135,000 – 115,000 years ago, the Sahara was green and they traveled across it, up the Nile River and into the Levant (Middle East). This was the first migration of modern man out of Africa. Oppenheimer

71

informs us that **African Homo sapiens "were fully modern, singing, dancing, painting humans long before they came out of their home continent."**

However, their northern migration was halted by one of the spectacular climate changes that rocked the world (probably because of irregular oscillations of the planet) at about this time. The humans set out in a beautiful interglacial period, but were probably brought to the verge of extinction in the ice age that occurred about 120,000 years ago. It is a wonder that they ever considered a second emigration! The survivors probably withdrew to their African home and bided their time (for a period of about 30,000 years !) as they recouped their energy and reproduced their numbers.

90.000 – 85,000 years ago the spirit of adventure took hold of their imaginations again. Homo erectus had experienced the same wanderlust. The humans crossed the Red Sea and traveled along the south coast of the Arabian peninsular.   Stone acheulean tools that they left behind have been found mixed with shells in shell-mounds (middens) are evidence of the fact that humans had found that sea food was a good and easily-obtained source of the protein they needed. They had become beach combers moving along quickly, always in search of seafood. Oppenheimer, the geneticist, points out that **all non-African people are descendants of this group of adventurers.**

85,000 – 75,000 years ago, the journey continued along the Indian Ocean coast until they reached West Indonesia. From there they moved to South China. The population no doubt grew immensely during the 10,000 years they occupied the region we now call Indo China. Little did they realize that they were about to be almost annihilated by one of the greatest volcanic eruptions to have occurred over the previous two million years.

74,000 years ago, Mt Toba in Sumatra erupted. Its date is established beyond the shadow of a doubt. So great was the eruption that the whole of India was covered with volcanic ash from one meter to five meters in depth. This gave rise to a six year "nuclear winter" and a 1,000 year ice age. The human population dropped to something like 10,000, a fact that is mirrored in the genetic map of human history.

74,000 – 65,000 years ago, the human race picked itself up and began the long march to the restoration of its population.. Groups passed by boat from Timor to Australia and from Borneo to New Guinea. During the ice-age Australia was much closer to Timor than it is now. In ice-age periods, the sea levels would drop as much as 150 meters.

65,000 – 52,000 years ago. People were retracing their footsteps back in a westerly direction. One route may have been up the Indus into Kashmir and on to Central Asia to European Russia and the Czech Republic and Germany. The climate began to warm 52,000 years ago and groups moved up the fertile crescent (Iraq), returning to the Levant. It is thought that, from there they crossed the Bosporus into Europe.

52,000 – 45,000 years ago. Another mini ice age occurred. Upper Peleolithic culture, marked by Aurignacian cultural remains. Cf. Wikipedia (The people of the Aurignacian culture produced worked bone points with grooves cut in the bottom, and some of the earliest known cave art, such as the animal engravings at Aldène in south-west France. Their flint tools were more varied than those of earlier cultural periods, employing finer blades struck from prepared cores rather than using crude flakes. The people also made pendants, bracelets and ivory beads, and three-dimensional figurines to ornament themselves.) Traces of this culture show that the Humans moved up the Danube and across to Austria and Hungary.

45,000 – 40,000 years ago, the new people entered Europe through the Levant and had reached as far as the Pyrenees and Spain. Encounters with the Neanderthals would have been frequent, but after about 10,000 years had passed, the Neanderthals went extinct and the only remaining Europeans were all Modern Humans.

So much for the return to the west from the Far East and the occupation of Europe. The time spans look small when put on a geological scale, but compared to our present A.D. era, they are still immense.

While Europe was becoming "humanized" by modern men no less intelligent than ourselves, groups of humans had been moving in various directions throughout Central Europe, Western Asia, China, Japan, North America and, finally, South America. The Bering Sea crossing to the New World took place about 22,000 years ago. An ice age depopulated the new race, but it survived and spread into South America about 17,000 years ago.

10,000 – 8,000 years ago, the prevailing ice age collapsed and the dawn of agriculture was ushered in. The Sahara was savannah.

74

again. Cf. The life-sized petrogryphs of giraffes in the Niger Region.

The ones from whom we may have descended or who were related to us by common ancestry had all become extinct. Our nearest living relatives are the chimpanzees and all intermediate species (The Australopithines) had become extinct. However, we had diverged from the chimpanzees over six million years ago. They are now of a different kind to ourselves. In other words, we are the last ones in the evolutionary adventure left standing.

Our next step will be to take a closer look at ourselves as the creatures standing at the top of the evolutionary ladder.

Chapter 10

The Explosion of Creativity

Animals such as chimpanzees get their name in the paper when they manage to do things that are taken for granted among humans, e.g. using two sticks to reach a banana. However, when it comes down to the question of whether they are intelligent in the human sense, even their greatest backers have to admit that they aren't. They differ from us not in degree, but in kind. [53] There are some very interesting pages in Ian Tattersall's book, *Becoming Human,* that illustrate the real difference between seemingly

---

[53] *Cf. Becoming Human* by Ian Tattersall P.30 –77 Ch.2 *Brain and Intelligence*

intelligent animals and ourselves. **A clear indication of this is found in the records of the tools used by men.**

The first fossils of modern humans were found in 1868, not in Africa, but in a rock shelter near Cro-Magnon in S.W.France. The fossils have been dated to 35,000 years ago. Similar finds were frequently made in other parts of Europe and it soon became obvious that we were looking at the first traces of modern humans in Europe.

They were called the **Cro-Magnon people**. They had obviously entered Europe about 40,000 years ago, but where had they come from? The answer is supplied in the previous chapter. They had come from Africa via Indonesia in a tumultuous human journey lasting 100,000 years. The tools they used set them apart from their predecessors. **The Cro-Magnon people were essentially no different from ourselves.**

About 2 million years ago Homo habilis had gained recognition for using stone tools of such primitiveness that it was hard at first to call them tools. These **Oldowan tools** found mainly in East Africa were improved on by later members of the Homo genus. The improved tools had come into use about 1.2 million years ago. They were hand-held, tear-drop shaped stones. Actually, the first fossil samples had been found at a place called Saint Acheul in France, which is why they are called **Acheulian tools**. These tools were the dominant ones used over the vast stretches of time of pre-modern-human history and the numbers of samples found in East Africa indicates that this is where they originated. By 500,000 years ago, Acheulian technology had found its way to Europe, probably by Homo Heidelbergensis and had been used there especially by Homo

Neanderthalensis during the 200,000 years they had existed.

Paleontologists have devoted a wide section in their science to the study of stone tools. We wont go into it here except to point out that one field of their studies is the Late Ice Age in Europe. (from 40,000 years ago to 10,000 years ago.) The Cro Magnon people gradually refined their tool-making skills, and experts divide the period into further periods marked by degrees in their unique tool-manufacturing skills. At first, we encountered the apparently intelligent use of stone tools by H.habilis, such as we might find in a Sea Otter's using stones to break open shell fish. This went on with exceedingly gradual sophistication over thousands of years, until Homo sapiens came into the picture and began to make tools to fit every occasion, just as we do today.

Seen from our 21st century stage of technical development, our Stone Age ancestors seem fairly primitive and slow-moving. However, this phenomenon is not surprising. The author's lifetime has seen huge advances in, for example, the way people wash their clothes – from boiling them and hand washing them on wash-boards, improved by the washing machine and mangle and now done automatically in front-loading Toshibas that change their cycles to the spin-dryer mode to finish things off! All progress takes time!

However, a closer look reveals that something extraordinary had happened. Homo sapiens doesn't have a lot of knowledge, but he has the ability to gain it. He demonstrates innovation and inventiveness and the will to apply these creative powers to the needs he experiences in his life. Animals don't have this inventiveness and even the Neanderthals, so close to modern man, gave no evidence of innovation. They just continued in the ways they had inherited from their forebears.

However, amongst the Cro Magnons, along with tools, art flourished, which has been conserved especially in the caves in Southern France, such as the Lascaux Caves. In his usual perceptive way Ian Tattersall describes these art forms, [54] mentioning also how Homo sapiens also confronted the mystery of death with religious ritual(Grave goods). He even had musical instruments and, of course, language.[55] Language consists of words that are symbols of ideas in our minds. The ideas can be organized to induce further understanding and to communicate our ideas to others. This symbolism is peculiar to our species and would have played a large part in the "creative explosion". **In a word, it is we modern humans that had arrived.**

To gain a deeper realization of this evolutionary step, I would advise the reader to open the Google page " The Creative Explosion". It introduces quite a few books born out of amazement at the coming into existence of Homo sapiens. E.g. ***Shell Beads and Behavioral Modernity*** by Kris Hirst. The book is introduced with the following words:. " "Traditionally, the transition between Middle and Upper Paleolithic is thought to have been marked at ~5,000 years ago. This transition was known as the "Creative Explosion", when Paleolithic cave paintings, sophisticated hunting techniques, portable art such as Venus figurines and other cultural traits were thought to have first appeared. But twenty years of research in South

---

[54] Ian Tattersall. *Becoming Human p.5-p.29 &p.173-p.187*

[55] Google: *The Origins of Language*

Africa, North Africa and the Levant have increased the time depth of this explosion, leading researchers to the inescapable conclusion that the "explosion" started in Africa, perhaps as long ago as 200,000 years ago, culminating in the Upper Paleolithic starting about 40,000 years ago." (In a cave in South Africa at Pinnacle Point, a team from the University of Arizona, found evidence that 164,000 years ago, people were eating shellfish, making complex tools, and using red ochre pigment – all modern man behaviors.[56])   In other words, the Cro-Magnons were natural descendants of Homo sapiens, who, 200,000 years previously, were already "talking, painting, singing adventurers", who had embarked on a unbelievable journey to the Far East, had suffered immensely, but had survived to discover a new home in Europe.

A further work already mentioned here is the book. ***Becoming Human*** by Ian Tattersall, a paleontologist and a curator of the American Museum of Natural History. The book is introduced by the words: " Human beings, in all their uniqueness, are the result of a long evolutionary process; and it is this which will be the central subject of this book. But since we started in Ice Age France, let's begin our evolutionary journey at the near end, so to speak, with a look at the astonishing record left by the Europeans of the late Ice Age, for these people provide us with **the earliest good record of the unique human capacity, fully formed:** evidence for what the science writer John Pfeiffer has called **"the creative explosion."** Not that this was an indigenous development; Europe was, until

---

[56] Smithsonian Com. *The Great Human Migration.*

about forty thousand years   previously, inhabited only by the Neanderthals: a distinctive and now-extinct group of humans belonging to the species Homo neanderthalensis. The Neanderthals were complex beings and talented users of the landscape they lived in: a far cry, indeed, from the brutish image with which generations of cartoonists have endowed them. But they left no evidence of the creative, innovative spark that is so conspicuous a characteristic of our own kind; and they were quite rapidly displaced by the first European Homo sapiens (Cro-Magnons), **who arrived at that time fully equipped with modern behaviors**."

Tattersall calls them **"our intellectual equals"** naturally capable of using "symbolism" (words standing as stymbols for ideas) a phenomenon that " lies at the very heart of what it means to be human."

The next question that comes to our minds is how to explain the event of "intelligence" exponentially superior to that of other species on the evolutionary ladder. Paleontologists have found that increase in the ability to use stone tools was accompanied by an increase in brain volume. One approach Darwin used in his study of evolution was that of embryology.   E.g. He finds in mammals the vestigial brain of reptiles now serving a different species etc.

Brain development had been a gradual process, but that final step that coincides with the existence of abstract ideas, innovation and inventiveness, awareness of various factors in the natural world (e.g. the calendar of the moon) and all the characteristic marks of what we term "human consciousness" of our place in the world can't be explained biologically. However, parts of the brain have taken up different relationships with other parts. In proportion to body size, our cortex is larger and more complicated than that of apes. The prefrontal cortex where thought processes are evidenced is better developed in humans.

Doesn't this indicate that "intelligence" might have a biological explanation? Craig Venter would seem to say that it does. He writes, "While we share most of our senses with the rest of the animal world, we have a most unique and exciting evolutionary development--our brain. It provides us with the ability to think, to reason, to predict and ponder the future. It enables us to ask questions and gives us the extraordinary capability to take over our own evolution by building complex tools that extend human capabilities millions of times further than would happen even with another billion years of Darwinian evolution." [57] However, Craig Venter, while listing the extraordinary effects of intelligence, still hasn't explained the origin of intelligence. Based on metaphysics, our contention is that the development of intelligence is beyond the power of physical causes. We could say that the presence of intelligence in man would have called for certain physical adjustments in the human brain to make its interface with intelligence smoother.

---

[57] Google: Craig Venter unveils synthetic life.

Chapter 11

The Origin of Human Intelligence

The next question to ask is where intelligence comes from. In replying to that question our society divides into three camps: 1) Materialists. 2) Theologians. 3) Metaphysicians.

J.Craig Venter is a brilliant researcher in biology whom Time Magazine in 2007 placed on the list of the world's most influential people. His team published the first private, complete (six-billion-letter) genome of an individual human (Venter's own DNA sequence.) When asked about his ideas on intelligence, he replied that he thought that a true scientist could not believe in supernatural explanations. In other words, as a scientist, he would feel obliged to search for some sort of biological explanation of the existence of the intelligence that characterizes ourselves and himself. His answer is fair enough. He refuses to give up despite the fact that no scientific answers to the question have been supplied so far. As a scientist, he cannot rest until phenomena have been explained by physical cause and effect.

In contrast to this outlook we have the belief of millions of Jews, Muslims and Christians in the Bible. Among them are fundamentalists whose literal interpretation of the Bible account of creation often flies in the face of scientific evidence. However, these are counterbalanced by the huge majority of believers who look on the Book of Genesis as a story stressing in poetic language that God is the Creator of all and that He breathed into the clay of man a spiritual soul. This is the constituting principle of the human being. Many scientists hold fast to this faith while, at the same time, exploring scientifically the biological and material phenomena in which this drama has unfolded.

Michael Angelo's Depiction of the Creation of Man

Then come the philosophers who rely on **metaphysical thinking.** "Metaphysics" means "Beyond Physics." It is with this camp that the author would like to dwell at length because their contentions can gain the assent of rational minds more thoroughly than some of the scientific hypotheses that we encounter.

Among the numerous books he wrote, the French metaphysician, Jacques Maritain, wrote one titled *" The Degrees of Knowledge."* It points out that physics includes in its object both the qualities and quantities involved in material things. Sciences such as geology and biology would be included under physics. Mathematics focuses on quantitative realities only and reasons to conclusions that are dictated by mathematical laws. Metaphysics abstracts from qualities and quantities and focuses on the very existence of things.

It is the most abstract of all the sciences, but its followers claim for it authority to discountenance all scientific conclusions that run counter to it, because its focus is on the laws that govern being itself. Its principles are easy to accept by common sense.

Apropos of this, we might mention Einstein's remarks when, in 1929, the astronomer Edwin Hubble published his findings that   proved the universe was expanding. The theory of *The Big Bang* holds that the galaxies are fleeing from one another like the sparks we see in huge displays of fireworks, and this is what appealed to Einstein. His mathematics had led him to this same conclusion, but it seemed so unlikely that he had tried to modify his mathematics to fit into the idea of a steady-state universe. He regarded this "subterfuge" as the biggest blunder of his career and he called the notion of the expanding universe, "The most beautiful and satisfactory explanation of the creation of the universe that I have ever heard." [58] In other words, mathematics, Maritain's second degree of abstraction, should have taken precedence over the physics (first degree of abstraction including visible qualities) that seemed to point to a steady-state universe.

In the same way, metaphysics would take precedence over any physical or mathematical theory that might postulate theories incompatible with it. For instance, one law of metaphysics would be that every effect demands an equivalent or sufficient cause. To postulate that intelligence is the result of biological causes would, according to metaphysicians, be ruled out because material

---

[58] Google. Cosmic fingerprints com.

(physical) causes would not be sufficient to produce immaterial effects such as conscious reflection , free will/morality and universal ideas.

At this point we might be asked to further elucidate why this might be so.

Actually, it might be necessary to point out here that "materialiam" doesn't allow for the existence of "spirit", but we contend that man can be explained only if we recognize in him an immaterial or non-biological principle of existence. The same refusal to recognize the spiritual side of man naturally extends to the denial of the existence of God, the Supreme Spirit.

Hence, **this chapter marks what could be termed a parting of the ways between materialists and theists.** Because of this, glimpses of the face of God in the background of the following pages can't be concealed.

We start our search for the truth of things as follows:

The first area we might explore to prove that intelligence demands immateriality or spirituality in man would be "**conscious reflection.**" How can a man look at himself as if from the outside. An animal knows things, but fails to realize that it is the one doing the knowing. It is conscious of things, but doesn't sufficiently reflect on itself in relation to the things it knows. For this reason animals don't make much progress in their way of living.

Richard Dawkins solves this problem very conveniently,[59] "I am not philosopher enough to discuss what (reflective) consciousness means." However, later on in the book, he introduces the "inner eye," a concept propounded by the psychologist Nicholas Humphrey to explain social skills brought about by knowing the psychological states of others in society by first knowing one's own inner psyche. "The inner eye is the evolved social-psychological organ," just as the outer eye is the visual organ. Dawkins then states, " So far, I find Humphrey's reasoning convincing."

Ian Tattersall, the author of the exceptionally interesting, thoughtful and profoundly researched book," *Becoming Human"* (p.192 ff) is also at a loss to explain reflective consciousness. He says, " **The properties of the human brain are "emergent" (i.e. seem to have come into existence from nowhere) and the mechanisms that lie behind these emergent properties remain among the most important unanswered questions in science."** "Neurobiologists and psychologists have done nothing to help us understand the quality we call (reflective) consciousness. Maybe, we should change the question and ask not what consciousness is but what it is for. Nicholas Humphrey calls it "the inner-eye", a unique property of the mind, based on "who-knows-what" (Tattersall's words) structural or physiochemical attributes, that allows the brain to observe itself at work"….It can be called 'self-reflexive insight.' Tattersall also concludes, " I must admit that I find Humphrey's

---

[59] *The Selfish Gene*   p.50 & p.280

metaphor persuasive, even though it doesn't embrace every aspect that common sense might attribute to consciousness."

Considering these statements by two well-known scientists, it is obvious that there is little acceptance of claims by metaphysics that reflective consciousness is evidence of an immaterial element (**spirit**) in man's make-up that couldn't have evolved from matter. Instead, a very scientific-sounding but gratuitous hypothesis is given precedence. The "unraveling of the neural code" in DNA (They say it might take years to do so) will reveal the existence or non-existence of the "inner eye" and the ultimate conclusion could be that the human soul simply doesn't exist, and that consciousness of being " other" and capable of looking at ourselves as though from the outside is a delusion !!

Linked with reflective consciousness is the experience that we have **free will**. We can be coerced into doing certain things by physical force(torture), by mind-altering drugs (their effect wears off with time), or by threats from blackmailers and difficult life situations, but, in spite of coercion, we are conscious that, at the core of our being, we have freedom of choice. We can say "Yes" to our torturers, but, at the same time, be saying "No" in our innermost being. There is something within us, in the core of our being, that is inviolable and beyond the reach of material force. This would be impossible unless there is something completely immaterial in our make-up that is immune to physical or material forces, our spirit.[60]

---

[60] *Signature in the Cell* by Stephen Meyer p.340 " For those of us trained in natural sciences, appeals to the mental realm sound

Another way of thinking could be based on **the claims of materialists themselves** that our minds function due to electrical neural impulses and the biochemical relay of signals throughout the nervous system of our bodies. If our thoughts are <u>purely</u> the products of these physical realities in our nervous system, it follows that we are not responsible for our actions, but we <u>are</u> conscious that we <u>are</u> responsible. We regard ourselves as being bound by moral considerations that often require us to go against our seemingly natural inclinations. In other words, we find that there is something within us that can over-rule the physical impulses that issue from the "biochemical relay of signals throughout our nervous system". The German Philosopher, Immanuel Kant (d.1804), held that we are governed by an ever-binding moral law of conscience, which he called a moral "categorical imperative." He thus indicates something in man above the material or physical elements in his make-up.

The second area where metaphysics claims to prove the existence of an immaterial spirit in man is the area of **universal ideas**.

The fact that we have **universal ideas** implies that we can

---

perilously vague, immeasurable and unscientific. We reflexively discount knowledge about what minds can do derived from introspection and ordinary experience and we, instead, credit only what we have learned through experimental studies. However, dismissing evidence from common experience can be a mistake...... Common experience often does count, but for modern scientific sensibilities, experimental results will always seem weightier."

abstract the essence of things e.g. "dog". It can be applied (universally) to every individual in a genus from a Great Dane to a Pocket Chihuahua without losing any of its exactitude, and conclusions about it can be reached without any sensory particularized representation of the individuals involved. e.g. "The dog is man's best friend."

Our ability to abstract universal ideas indicates the presence of something within us that relies on knowledge of things obtained through the senses, but which abstracts from their individuating characteristics. One principle in psychology as seen in metaphysics is " There is nothing in our minds that doesn't come from our senses." This ability to abstract the essence of what is presented by the senses is called "the active faculty of thought or intellect". It uses sense knowledge as grist for its mill and produces ideas or thoughts that can be predicated of all things (universally) that *share* the essence that has been abstracted. It can also formulate the connectedness between different universal ideas. This way of knowing is peculiar to human beings. It defines them and, at the same time, accounts for the limitations in their knowledge.

We can have ideas of realities that exist as such only in our world of universals e.g. truth, honesty, justice, evil, cause and effect etc, but manifestations of these ideas exist in particularized situations before they are abstracted and then received into our intelligence as universal ideas. There is something working in us in addition to our sense faculties. **Our senses are cognitive faculties** that grasp individual things as individuals. Any of our senses are themselves material e.g. the eye or the ear are physical organs, and the end-product of their actions is neurological stimulation ending in sight or hearing. On the other hand, **the human intellect is a cognitive faculty** that doesn't grasp things as individuals, but as essences on which individuals are patterned and represents them in ideas that can be applied universally. Such a faculty could not itself be material, because its end-product doesn't exist in a material manner. Sight or hearing can be studied under a microscope in the sense organs that produce them. A universal idea is one that has no physical existence comparable to the psycho/neurological existence stimulated by our senses. .

There is a dictum, that everything operates according to the sort of being it is; that is, its operation is in accordance with its nature. Thus we must conclude that there is something in man that must be immaterial because it is not constrained by matter in its operations. This puts it above matter, but it is, at the same time, somehow contiguous with matter and communicates its existence to matter. It is the form (soul) of a man's body." The body and the soul are the two principles that function to produce man.

In the case of human beings, body and soul constitute one being (man) and all human activities are brought about by the function of body and soul together. Intellection requires sensation because universal ideas are abstracted from individual objects that are perceived by the senses. Hence, it must be said that, if human beings are to act as human beings, they must have a body as well as an intellect.

By expressing it in this way, it is hoped that any misunderstanding leading to dualism might be avoided. Dualism would regard the soul as a separate being here and now from the body. A condition comparable to a charioteer (soul) in a chariot (body). <u>Body and spirit are not separate beings, but are two principles of the one being.</u> As Tattersall pointed out, man is an "emergent" being. The metaphysicians say that the reason for his having "emerged" is that **somewhere along the evolutionary journey the body was ready to be informed by spirit, resulting in a new kind of being which we call human.**

To keep on insisting that corporeal matter will one day yield up its secrets and that biological discoveries about the brain will be able to explain human intelligence is, according to metaphysics, to take a road that leads to nowhere. What metaphysicians claim is that there is an immaterial dimension in man that is "emergent," and completely unexplainable biologically.

The words of Ian Tattersall that I have already quoted might be worth taking a second look at. He says,[61] **" The properties of the human brain are 'emergent' (seem to have come into existence from nowhere), and the mechanisms that lie behind these emergent properties remain among the most important unanswered questions in science."** He still regards intellectual understanding as a "property of the human brain", but he calls the properties "emergent". Couldn't it be that they are emergent because they owe their existence to an immaterial formative cause outside the human brain?

---

[61] P.192 ff, *Becoming Human*

Materialists try to discover what makes an organism change and it has come up with the extraordinary discovery of the DNA code as outlined in a previous chapter. They hypothesize that even the intellect is a property dependent on DNA. While science tries to prove this hypothesis, metaphysics looks only at the existence of reflective consciousness, free will, moral obligation and universal ideas and makes the claim that the formative principle of man's nature is immaterial despite its being intricately associated with the body (physical brain) in its natural operation.

Spirituality in Man

Biology can't explain reflective consciousness. In the primates the brain had developed to an extraordinary state of awareness, but, just as the eye cannot see itself, reflection on the self as an objector as something known from the outside was beyond the power of anything inner in the brain itself. The ability to know the self, as it were from the outside, implies the existence of a power independent of the physical brain. This power must therefore be rooted in an immaterial element working in tandem with the physical brain, but extending beyond it in its reach.

In other words, immateriality or spirit must have been created in the Primates to form a new type of being – a man – or a spiritualized Primate. The creation of spirit was a gift from God, the pure spirit and Creator of the universe. It was the necessary step in God's plan to raise up the material world through the material creature who had evolved over the eons.

This step could be called God-assisted evolution in man. A corollory would be that, since spirit or soul has no biological cause,

God is directly involved in the coming into existence of each and every human person. A fact that shows the closeness to God of the parents in their procreation of children. It also shows the dignity of man.

Another corollary would be that there seems to be nothing to prevent the soul of man surviving the dissolution or the death of the body. However, the soul without the body, whose formative cause it is, is hard to imagine. For an explanation of this mystery, the next book in this series, *The Best of Life,* is recommended. It treats not the survival of the soul after death, but the resurrection of the soul and body in their essential unity.

Immediately after the creation of spirit and intelligence, things would have gone as as usual at first. However, the vital flux in the primates had manifested itself in superb awareness through the senses of concrete, individual situations. However, in man, the vital flux manifested itself by its command of situations involving the person and its relationship with everything else in the world. We can quote here the observation made by the Jesuit paleontologist Pierre Teilhard de Chardin (1881-1955) "According to all appearances, life went on in man, as in other animals, after the threshold of thought had been reached, just as it had always done, as though nothing had happened.... But, like a river enriched by flowing through an alluvial plain, the vital flux, as it crossed the stages of reflection, was charged with new principles and, as a result, manifested new activities." (p.174)

The difference between the Primates and Man was characterized by an ever-widening gap in their respective grasps of reality. Man with his reflective consciousness and his ability to form and relate universal ideas gave him a huge advantasge over the Primates.

The first human beings

As regards **the precise time of the birth of man**, there are no scholars who would deny that Homo sapiens(Modern Man) had come to birth by about 200,000 years ago. De Chardin thinks that evolution marked by reflection could have come about much earlier, somewhere about 1,000,000 years ago, represented by Homo pithecanthropus (who also goes under the names of Java Man and Homo erectus) and Sinanthropus (Peking Man or Sinanthropus erectus, 500,000 y.a.). These "men" probably had primitive speech, they worked stones and possibly made fire, activities on the level of reflection. However, they were not "precisely us." (p.196, 198).

They might have been already, in the full sense of the word, intelligent beings." (p.195).They would have been exploring their outer and inner world in ways different to those that the primates used and would have been experimenting with how to use their bodies in ways expressive of their intelligence.

de Chardin points out that " Man came into the world silently. He trod so softly that, when we catch sight of him, as revealed by those indestructible stone instruments, we find him

sprawled all over the world from the Cape of Good Hope to Peking."
(p.186). For paleontologists, the first man is already a crowd and his
infancy is made up of thousands of years." (p.186) The face of the
first man will, like the origin of life, remain forever beyond our
grasp.(p.170).

In the end, from the organic point of view, the whole
metamorphosis leading to man depends on the question of a better
brain, allowing primitive man to reach his potential under the
influence of his newly created spiritual nature. [62]

Chapter 12

Is there Evolution Still to Come?

In order not to be side-tracked from the main purpose of
this study, I have avoided any mention of genetics being used in
Medicine. E.g. Certain bacteria and viruses develop immunity to
medications. This is actually a case of almost instant evolution on the
part of the bacteria concerned. Darwin, who held that long
time-spans were needed for evolution, would have liked to know
this.

I have also refrained from mentioning what are called
mitochondrial DNA and y-chromosomes, which have been used in

---

[62] The page numbers in the last few paragraphs are those of de Chardin's *The
Phenomenon of Man*. (Harpur Colophon Edition, published in 1975.)

paleontology to trace the history of people back to their maternal or paternal common ancestors. This has been a valuable instrument for dating fossils and pin-pointing common ancestors, but it is outside the scope of this booklet.

However, I will quote here a snippet from a recent copy of the Japan Times (29/6/2010). It is headed " Diabetes gene search finds new links. An international team of scientists, working on the largest study to date, looking at DNA and Type 2 diabetes, say they have found twelve new gene links that offer important clues to how the chronic disease works."etc.

The above paragraph shows that evolution is no longer a matter of "selfish genes" going it alone. They are being interfered with by Homo sapiens.

Richard Dawkins has coined a new word "meme" (mimicry) to rhyme with "gene" and to indicate that we have to " throw out the gene as the sole basis of our ideas about evolution. I am an enthusiastic Darwinist, but I think Darwinism is too big a theory to be confined to the narrow context of the gene, and the law that all life evolves by the differential survival of replicating entities…………I think that a new kind of replicator has emerged on the planet. It is staring us in the face, but it is still in its infancy, still drifting clumsily about in the primeval soup. However, it is already achieving evolutionary change at a rate that leaves the old gene panting far behind it.

The new soup is the soup of human culture and the unit of cultural transmission is a unit of imitation. I call this the "meme".

Dawkins then depicts ideas jumping from brain to brain by

imitation. As examples he cites "belief in life after death" and "belief in the existence of God." These ideas are aided by the spoken and written word, by sublime music and by exquisite paintings.

He asks what it is in the idea of "God" that penetrates the cultural environment so thoroughly? He answers the question: " It is its great psychological appeal. It provides a superficially plausible answer to deep and troubling questions about existence. It suggests that injustices will be righted, and the "everlasting arms" hold out a curtain to cushion us against our inadequacy." (Author's note: Some might object and claim that the acceptance of God's existence relies on reasoning removed from such emotional appeal as that given here.)

Dawkins continues: "God is a meme with high survival value. I have emphasized[63] that we mustn't think of genes as conscious, purposeful agents, but blind natural selection makes them behave as if they are purposeful. The same can be said of memes. In neither case must we get mystical about it. In both cases, "purpose" is only a metaphor…….." Faith also is part of the religious meme complex. It means blind trust in the absence of evidence, even in the teeth of evidence. For faith, nothing is more lethal than a demand for evidence. The meme for faith secures its own perpetuation by discouraging rational enquiry. Once, the genes built "survival machines" (biological entities). With "brains, " the memes immediately took over…… We are built as gene machines and cultured as meme machines."

---

[63] *The Selfish Gene* Richard Dawkins. (p.196. 2006 Edition)

Dawkins is a writer easy to read, and reading him is a pleasure. He has such <u>blind trust in the absence of evidence</u> for gratuitous statements he makes with innocent aplomb! However, he has a point. With the advent of intelligence, man is no longer completely at the mercy of his genes, and human culture will develop (evolve) to a great extent under the guidance of intelligence. Natural genetic–driven evolution will still exist, however, and researchers have found that, in the last 5.000 years, it has speeded up.[64] They have found evidence of recent selection in 7% of all human genes, including lighter skin and blue eyes in northern Europe and partial resistance to diseases, such as malaria, among some African populations. Environmental change, especially what we eat, is one reason for these changes.

---

[64] Google: BBC NEWS | Science/Nature | **Human evolution is 'speeding up'**

Chapter 13

"The 'Science' versus 'Intelligent Design' Controversy.

Recently [65] there was a statement, " Intelligent Design to be taught in Queensland Schools under the National Curriculum." The article stirred up a hornet's nest of opposition expressed on the web.

Before we go ahead and consider the pros and cons of Intelligent Design, we should define the words that are used. **"Creationism"** is the religious belief that humanity, all life on the earth and the universe are direct creations by God. Creationism claimed that scientific theories such as evolution, which are based on observations of processes in nature, can't account for the history, diversity and complexity of life on earth.

Strict creationists usually base their beliefs on a literal reading of the Genesis creation narrative. In the 1920's, the creationists in America succeeded in having teaching about evolution banned from the school curriculum.

Reactions against this in the mid 1960's resulted in Creationism itself being banned from the curriculum, on the grounds of America's law about the separation of church and state, which forbids teaching religion in public schools.

With this restriction in mind the creationists stripped their

---

[65]   Cf. Courier Mail, 10thJune 2010.

texts about creationism of overt Biblical references and they renamed their teaching **" Creation Science".** This also was outlawed in 1987 for being "Creationism in disguise".

The next attempt to counter the proliferating atheism in society was a theory called "**Intelligent Design**", which goes under the letters " ID". It was not just a rehash of "Creation Science". As we shall explain more fully later in this chapter, it was an attempt to use scientific arguments to disprove the claims of materialist evolutionists and is **not to be confused with the proof for the existence of God based on the design we observe in nature**.

When I was a high school student in a Catholic school in Queensland, Australia, we were taught in Religion Class reasons for belief in God. One book recommended to us students was written by a Doctor of Medicine, W.V.McEvoy, under the title of *Things in Me That Make Me a Believer.* The author writes about the intricate order and design of our various bodily organs and the connectedness of these entities of which our bodies are composed.

As a young student, I found this book very convincing. It is easy to believe in the existence of a Designer when we are confronted by such amazing design. However, since those days, science has shown that what was regarded as design produced directly by the Creator is produced by Darwinian selection and other processes. McEvoy's explanation went straight to the existence of an intelligent creator. However, in these times he might have pointed out that this exquisite design is directly because of extremely versatile and evolutionarily active DNA, and only indirectly because of the creativity of an intelligent Designer.

I think that when most people hear of the theory of intelligent design, they think of it in terms similar to what I have outlined above as one underpinning of my boyhood faith. This way of thinking would be fairly universal, but Darwinism has been perceived by some scientists as invalidating the argument from design (at least in the case of biological life) for the existence of God.

However, the theory called **"Intelligent Design"** (ID), written with capital letters, is not the same as the above description of my boyhood "proof" for the existence of God. It came into the spotlight in the early 1990's in a book, *Darwin on Trial,* written by Philip Johnson. He is a lawyer at the University of California at Berkley and was reacting against atheistic attempts to use Darwinism as a lever to exclude God from the milieu of American life. He wanted to present scientific arguments against those who held that science has made God superfluous. He was assisted in this by the biologist, Michael Behe, whose book *Darwin's Black Box* expounded the notion of **"irreducible complexity"** a theory that holds that even if evolution is admitted, there are some organisms that are " irreducibly complex" (which couldn't have evolved piecemeal). These complex entities couldn't have evolved without help from a supernatural source. A key example of such an organism proposed by Michael Behe [66] is the mammalian eye. However, evolutionists refuse to accept that the eye is too complex to have been the product of evolution. There are, in nature, organisms that are sensitive to

---

[66] Cf. Behe's book, *Darwin's Black Box*

light. Collins gives the example of the aquatic creature, Nautilus. It has a pin-hole cavity to admit light which assists the organism in direction finding. Also a transparent jelly-like substance covers light-sensitive cells in other organisms etc. Over millions of years small incremental adaptations could have resulted in the mammalian eye etc. Francis Collins points out that there is always the danger that what is not understood now may become understood hereafter. He is the enemy of any use of what he calls "The God of the Gaps," saying that, in the long run, when science does succeed in understanding physical realities not now understood, it brings Faith into disrepute, if believers have invoked divine intervention in default of the inability of science to explain the phenomena at the time. Collins treats the subject of this chapter from p.182 to p.195 in *The Language of God.*

Darwin conceded that, "If it could be demonstrated that any complex organism existed, which could not possibly have been formed by numerous, successive, slight modifications, my theory would absolutely break down."[67]

In other words, Darwin was convinced that his "natural selection of random advantageous genetic mutations" was a natural law that he had discovered. For him and his followers it is the way species have evolved and, to assert, as ID does, that God would need to intervene to supplement defects of this law would destroy the possibility of natural science. However, to accept Darwin's idea as a

---

[67] Charles Darwin, "On the Origin of Species by Means of Natural Selection) cf. Google Darwin's theory of evolution. A theory in crisis.

natural law is equivalent to accepting the hypothesis that God has entrusted the development of biological forms completely to "natural selection of random advantageous mutations." (This concept is explained more fully from p.13 of this book.)

ID was thus invented to assist Christian believers to defend their beliefs by scientific methods against the modern onslaught of atheism. However, despite its good intentions, ID has not been supported by many scientists, even among those who share the Christian Faith, but its proponents have been highly vocal in propagating the theory.

There is **another version of ID** which claims that messages transmitted by DNA in a cell must have originated through an intelligent agent. This theory is the basis of the ideas presented very convincingly by Stephen Meyer in his 2009 best seller, *The Signature in the Cell.* However, even this theory is also criticized by its opponents. It is possible that Meyer's ideas will be influential in persuading scientists that "information" in DNA is a **specified complexity** that demands an intelligent Cause for its existence. It stands at the point of origin of the force that governs biological life, just as the Big Bang stands at the point of origin of the universe.

The above paragraphs outline the background to the disputes about teaching Intelligent Design in schools. To this author, it seems that it would be a dangerous step to teach ID as an alternative to the prevailing science, without a very clear definition of terms. Teaching TE (Theistic Evolution. Cf.Foreword p.5)   would be a better alternative.)

The whole thing has already been argued out in America and other countries, so we have some precedents to learn from. I will quote a few sources here:

In the United States. In 2004, the Dover School Board in Pennsylvania voted that a statement be read to students in biology classes mentioning Intelligent Design (ID) in the universe. This was opposed by scientists and science teachers in the district and legal proceedings were launched. On Dec.20th 2005, Federal Judge John E. Jones ruled that the school board had violated the Constitution. He addressed the court as follows: " In making this determination, we have addressed the seminal question of whether ID is science. **We have concluded that it is not, and moreover that ID cannot uncouple itself from its creationist and, thus, its religious antecedents."**

This case mirrors a whole host of similar school board dilemmas in America, where the separation of religion and the State is part of the Constitution.

The Council of Europe: Oct.4, 2007, released a resolution that rejects the idea that creationism in any form, including ID, can be considered scientific, but recommends its inclusion in religion and culture classes.

The United Kingdom. The teaching of evolution is compulsory in publicly funded schools. The curriculum says," the fossil record is evidence for evolution and variation and selection may lead to evolution or extinction." The Archbishop of Canterbury, Dr Rowan Williams, leader of the Church of England, has expressed his view that creationism should not be taught in public schools,

"The government has stated that neither ID nor creationism are recognized scientific theories and they are not included in the science curriculum."

Australia vigorously supports the rights of teachers to teach science, including evolution, unhindered by religious restrictions.

The above statements are not about the existence or non-existence of God. They are only about whether conclusions of the theory called Intelligent Design (ID) can be called scientific conclusions. The American Judge, John E. Jones, stated: " We have concluded that **ID is not science** and can't uncouple itself from its creationist and religious antecedents." (Its invocation of supernatural intervention to explain irreducible complexity is an abdication of any claim to be scientific, and the key examples it uses of irreducible complexity have already been proved to be reducible. [68])

This gives rise to **the question of what science is**. A good definition is supplied by Google: "Science is the study, documentation and collection of evidence pertaining to observable and quantifiable, naturally occurring objects, phenomena and processes within the universe."

Firstly, it should be noted that many scientists have a bone to pick with the above definition of science, e.g. Stephen Meyer,[69]

---

[68] Collins p.187 ff

[69] E.g. Stephen C Meyer in *Signature in the Cell.*(Harper Collins , Publisher) He objects to the very restrictive definition of science as given here. He maintains that the existence of intelligible coding in the DNA is a well-known natural phenomenon that needs explanation. To rule

but it is accepted here because it is, at this point of time, the definition that has enjoyed a long history of acceptance, and was in the background of Judge Jones's understanding of "science."

It should also be noted that, in actual fact, a huge number of scientists accept the existence of God. However, they haven't been doing this as scientists, who are limited by the definition of science given above. They do it only when they change gears from scientific reasoning to the broader type of thinking that we use in everyday life. This type of thinking has a close affinity with metaphysical reasoning..

Teaching belief in God is not part of the science course, because conclusions that involve "Spirit" are not "testable", "observable" or " quantifiable" phenomena. In addition, it should be mentioned that God cannot be conceptualized, so, to use God as a "default" entity or "god of the gaps" to explain natural phenomena is unacceptable as far as scientists are concerned.

Science is limited by its very definition. Apropos of this

---

out the possibility of intelligent design of DNA, just because of the way "science" has been defined, is, for Meyer, a flaw in the definition of science because it prevents a full investigation of one of the most fundamental phenomena in nature.

In his thesis, which the Time Literary Supplement chose as the Book of the Year, in 2009, he eliminates all the other attempts at explaining the DNA code and invokes intelligent design as the only viable explanation. His reasons for his claims are very convincing and were pursued with the backing of many of the brightest scientists, mathematicians and investigators of the origins of life in Cambridge and American Universities" (He doesn't set out to prove that "God needed to intervene in cases of irreducible, intricate complexity found throughout the history of the human race", a claim that would label him as a proponent of the general theory of Intelligent Design (ID). He claimed only to have established that the existence of the DNA genetic code cannot be explained unless it is the product of intelligence.)

idea, Collins[70] quotes Stephen Jay Gould (Previous to his death in 2002, he was the most widely read proponent of evolution.) *" Science simply cannot by its legitimate methods adjudicate the issue of God's possible superintendence of nature. We neither affirm nor deny it. We simply cannot comment on it as scientists. "*

Another point that could be discussed is that which, in America, occasioned the verdict against a " statement being read to students in biology classes mentioning that Intelligent Design in the universe is an alternative to the purely materialist version taught in biology class." (Cf. P.66 of this text). It must be remembered that in America there are strict laws against teaching religious views in schools. Australia allows more latitude in this matter. Students studying biology and evolution could be taught that the failure of their biology course to mention God or the immateriality of the human soul is due to the fact that science refuses to accept intuitive conclusions and stops one step before ultimate or transcendental conclusions are reached. A suggested statement that could be helpful for students is presented as follows.

---

[70] *The Language of God* p.165

In your science course, you may be told that the Big Bang marks the origin of the universe, but there will be no further explanation of the nature of the cause of the Big Bang because it is not quantifiable or observable. In other words, it will appear that God as Creator is being ignored by science. If you want to think more about this, you will have to do it privately and out of class.

To help your thinking, you might make enquiries about how others have reached conclusions about God. For instance, Charles Darwin: " I am challenged by the extreme difficulty, or rather the impossibility, of conceiving this immense and wonderful universe, including man with his capacity for looking far backwards and far into the future, as the result of blind chance. When thus reflecting, I feel compelled to look to a First Cause having an intelligent mind in some degree analogous to that of man; and I deserve to be called a Theist."[71] (At one stage of his life, Darwin had claimed to be an agnostic.)

In Darwin's *The origin of the Species,* he wrote about his scientific conclusions, "There is a grandeur in this view of life (evolution), with its several powers, having been originally breathed by the Creator into a few forms or into one; whilst this planet has gone cycling on according to the fixed laws of gravity, from so simple a beginning. Endless forms most beautiful and most wonderful have been, and are being evolved." Darwin wrote these words in the final sentence of his book. He had formulated a scientific theory that evolutionary theories have been built on, but he

[71] Words from *Finding Darwin's God* by R.Miller(N.Y.HarperCollins,1999)p.287

stepped away from science to acknowledge God as the first cause of life.

You will have to reach your conclusion about God in the same way – by thinking about it. Things don't necessarily need to be proved by science before they can claim acceptance. There are other ways to reach valid conclusions, and they are used all the time in our everyday lives.

The laws of nature such as gravity and things like the DNA code are phenomena that are taken for granted. No effort will be made to link them with an Intelligent Creator. They will be studied and used, but science does not set itself the task of discovering the ultimate cause of their existence. The study of ultimate causes is the realm of metaphysics. You will have to make this discovery by yourself. It may help if you remember the words of Einstein, "the harmony of natural law reveals an Intelligence of such superiority that, compared with it, all the systematic thinking and acting of human beings is an utterly insignificant reflection..."

Finally, in the science class, you will be taught that studies about the human brain in the future will eventually explain how reflective consciousness, free-will and abstract thought are able to be explained. You will also have to do some thinking about this by yourself. You might find it helpful to read from p.82 of this booklet. However, you are conscious that there is something in you that material forces can't touch, you are free and this would be impossible if you were just a physical body. You realize that there is an immaterial element in your make-up because material force cannot "push you around." They can deprive you of your freedom of movement, but they can't deprive you of your freedom of choice in the inner depths of your mind.

Science limits itself to explaining physical phenomena. If it starts teaching about immaterial reality, it is (according to the definition given here) no longer science. A believer in God can accept the whole body of science, but goes that one step further and relies on intuition and everyday (metaphysical) reasoning to reach conclusions about God and the human soul that science baulks at.

Both photography and art can depict the same scene. Photography presents it just as it is, while art may go further and accent aspects of the scene related to such ideas as the ephemeral nature of life as exemplified by falling cherry blossoms. A photography teacher might refuse to teach art because he is not at home with such art forms as abstract art, impressionism, surrealism etc. Likewise, a science teacher would refuse to teach what I have called metaphysical conclusions, because it involves a different discipline and goes beyond the limits that science sets for itself. There are thousands of science teachers who believe in the existence of God and human spirituality, but who would not consider any need to teach these beliefs in their science classes and would regard any attempt to do so as outside the scope of their commission to teach the science course.